Introduction to Crop Production

JOHN DEERE

AGRICULTURAL PRIMER

Introduction to Crop Production
ISBN-0-86691-363-7

FP701NC (2015)   (ENGLISH)

A brief overview of the processes, principles, and practices
associated with crop production including soil management,
planting, harvesting and more.

Deere & Company
PRINTED IN U.S.A.
26OCT15

Introduction

To download the latest catalog or for order information please visit: www.johndeere.com/publications

DEERE & COMPANY

JOHN DEERE PUBLISHING

One John Deere Place

Moline, IL 61265

Agricultural Primer

Agricultural Primer (AP) is a series created by Deere & Company. Each book in the series is conceived, researched, outlined, edited, and published by Deere & Company, John Deere Publishing. Authors are selected to provide a basic technical manuscript that could be edited and rewritten by staff editors.

HOW TO USE THE MANUAL: This AP manual can be used by anyone — experienced mechanics, shop trainees, vocational students, and lay readers. The instructions are written in simple language so that they can be easily understood.

Persons not familiar with the topics discussed in this book should begin with Chapter 1 and then study the chapters in sequence. The experienced person can find what is needed on the "Contents" page.

Each guide was written by Deere & Company, John Deere Publishing staff in cooperation with the technical writers, illustrators, and editors at Almon, Inc. — a full-service technical publications company headquartered in Waukesha, Wisconsin (www.almoninc.com).

Introduction to Crop Production

Introduction to Crop Production is a brief overview of the processes, principles, and practices associated with crop production. A complete glossary of terms used within this book, related to crop production, is also included. This book discusses:

- Soil management
- Tillage
- Planting
- Chemical application management
- Combine harvesting
- Hay and forage harvesting
- Precision farming
- Machinery management

Acknowledgments

John Deere gratefully acknowledges the following people for their contributions to this manual:

Gerald Higdon, received an MBA in Technology Management from the University of Phoenix, and serves as Publisher for the John Deere Ag and Turf division at the Headquarters in Moline, IL.

Sue Gray, received her B.S. in Farm Management and M.S. in Crop Production from the University of Illinois. As an agronomic consultant, she has assisted John Deere as well as seed, chemical, and agricultural research firms in training, research, and product development.

John Deere also acknowledges the authors of previous John Deere publications including: Soil Management Solutions Handbook, Soil Management FBM11101NC, Tillage FMO11104NC, Planting FMO12103B, Chemical Applications Management FBM19103NC, Combine Harvesting FMO15105NC, Hay and Forage Harvesting FMO14105NC, The Precision-Farming Guide for Agriculturists FP403NC, and Machinery Management FBM17106NC. Portions of these publications along with updated content contributed to Introduction to Crop Production.

ISBN 0-86691-363-7

FP701NC

Preface

As recently as the 1940s, almost one quarter of the U.S. population lived on farms. Even as a large percentage of those residents moved from the farm and to other employment opportunities, a basic understanding of farm practices was still shared by many. Fifty years later, all but 1% of the U.S. had left the farm, and along with the rapid migration went the knowledge of the activities involved in growing their food.

Introduction to Crop Production is written for those with a desire to revisit our agricultural roots — but in the context of the very modern industry into which agriculture has grown. Resources to feed our ever-growing population are met by extremely efficient and productive individuals remaining on the farm. This efficiency and productivity has been possible due to technologies developed by many of the farmers' partners. Some of these technologies deal with:

- Seed — From hybrids to gene transfer, plant genetics have literally and figuratively provided the seed for rapid and continued growth in yields.
- Agricultural chemicals — From synthetic fertilizers to pest protection, with improved application techniques for both, the agricultural chemical industry has helped to increase production while minimizing impacts from outside threats.
- Equipment — From seed placement through crop care and right up to harvest, equipment has supported the boosts in productivity and yield while greatly reducing labor input, allowing the farmer to move from laborer to manager.
- Information — Adoption rates of emerging technologies have increased markedly as research and information can be rapidly accumulated and disseminated to support decision-making. In many cases, the information is specific to the farmer's own operation, as precision-farming capabilities support simultaneous collection, reporting, and recommendations specific to each field or area within it.

The objective of this book is to provide a brief overview of the processes, principles, and practices associated with production of the major crops grown in the United States. Some of the questions that this book addresses include:

- What are the basic needs of a growing crop, and how does the farmer accommodate them?
- What tools does the farmer employ to productively plant, nourish, protect, and harvest a crop?
- What drives the choices and decisions involved in managing crop production activities?

Crop production is a rewarding industry to understand and support; this publication will provide a basis for that growth to begin.

MM16633,000258D -19-10DEC10-1/1

Contents

Continued on next page

Original Instructions. All information, illustrations and specifications in this manual are based on the latest information available at the time of publication. The right is reserved to make changes at any time without notice.

Contents

Page

Machinery Management

Glossary of Terms

Crop Production Yesterday and Today

Introduction

As recently as the 1940s, almost one quarter of the U.S. population lived on farms. Even as a large percentage of those residents moved from the farm and to other employment opportunities, a basic understanding of farm practices was still shared by many. Fifty years later, all but 1% of the U.S. had left the farm, and along with the rapid migration went the knowledge of the activities involved in growing their food.

OUO1023,00040C2 -19-22OCT15-1/5

According to the U.S. Department of Agriculture, although the number of farmers is increasing slightly, they continue to shrink as a population compared to the 300-plus millon population base and growth trend of the United States.

	2007	2002	% Change
All Farm Operators	3,281,534	3,053,801	+7
Under 45 Years	732,322	851,091	-14
45 to 64 Years	1,725,777	1,527,742	+13
65 Years and Older	823,435	674,968	+22

Census of all farm operators

OUO1023,00040C2 -19-22OCT15-2/5

Not only are the numbers of farmers decreasing per capita, according to recent census data, the average age of the farmer is rapidly increasing. One might surmise that this trend would also lend toward less knowledge transfer as the population ages.

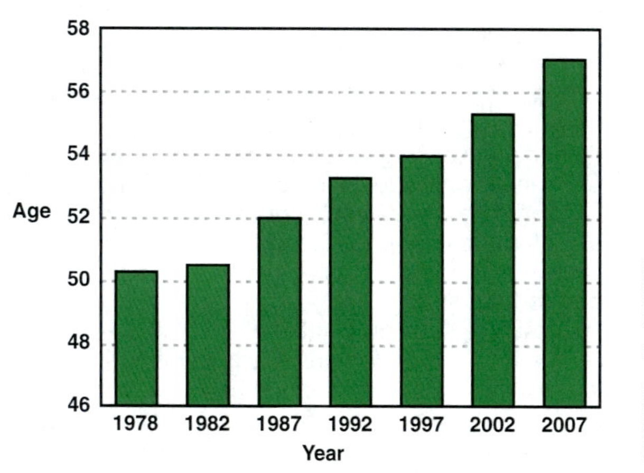

Average age of principal operator

Continued on next page

OUO1023,00040C2 -19-22OCT15-3/5

Along with the age factor and the number of farmers in business today, a broader population base is also impacting the scope of those who call themselves farmers today. According to the USDA: "The 2007 census shows that U.S. farm operators are becoming more diverse. Of the 2.2 million farms in the United States, 1.83 million have a white male principal operator. The number of principal operators of all races and ethnic backgrounds has increased 4% since 2002, but the growth in the number of non-white operators has outpaced this overall growth. The number of operators of Hispanic origin has also increased 10% since 2002." "One of the most significant changes in the 2007 Census of Agriculture is the increase in female farm operators, both in terms of the absolute number and the percentage of all principal operators. There were 306,209 female principal operators counted in 2007, up from 237,819 in 2002 – an increase of almost 30%. "

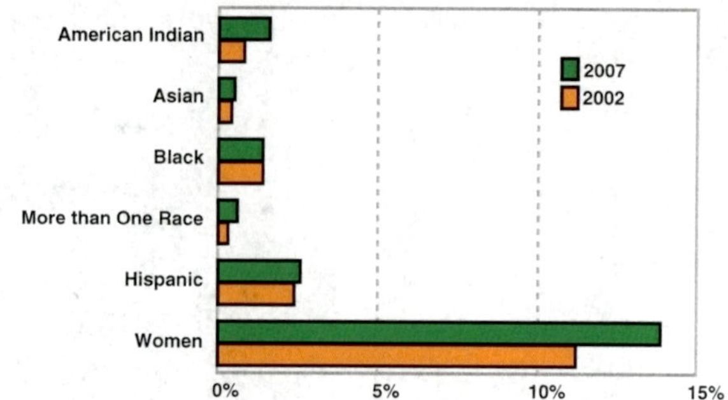

Percent of farm operators by race, ethnicity, and gender

Resources to feed our ever-growing population are met by extremely efficient and productive individuals remaining on the farm. This efficiency and productivity has been possible due to technologies developed by many of the farmers' partners. Some of these technologies deal with:

- Seed — From hybrids to gene transfer, plant genetics have literally and figuratively provided the seed for rapid and continued growth in yields.
- Agricultural chemicals — From synthetic fertilizers to pest protection, with improved application techniques for both, the agricultural chemical industry has helped to increase production while minimizing impacts from outside threats.
- Information — Adoption rates of emerging technologies have increased markedly as research and information can be rapidly accumulated and disseminated to support decision-making. In many cases, the information is specific to the farmer's own operation, as precision-farming capabilities support simultaneous collection, reporting, and recommendations specific to each field or area within it.
- Equipment — From seed placement through crop care and right up to harvest, equipment has supported the boosts in productivity and yield while greatly reducing labor input, allowing the farmer to move from laborer to manager.
One example, would be John Deere's CCS Seed Delivery Refuge Plus Option Planter. This planter was designed with a three tank configuration as a solution to changing seeding requirements. These changing requirements include the need to seed different varieties along with having versatility of configuration, the need to conform to Integrated Resource Management (IRM) guidelines and mandates associated with genetically modified organism (GMO) crops, and the need to maintain bulk seed productivity. The CCS Seed Delivery Refuge Plus Option Planter is just one way John Deere has met the expectations of the changing agriculture industry. The ever-changing crop production industry makes it more difficult than ever for unfamiliar persons to understand the basics of how a crop is grown.

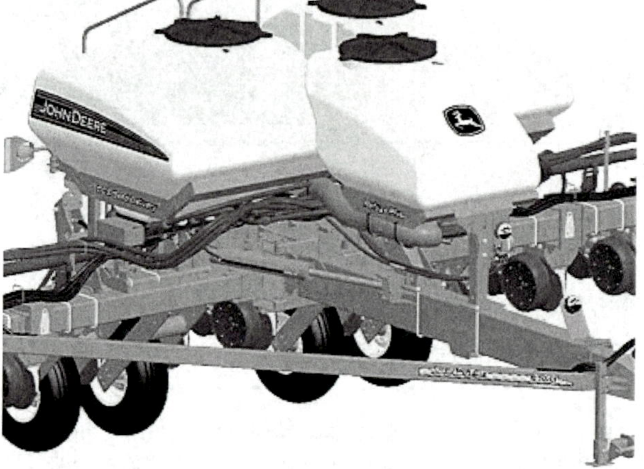

CCS Seed Delivery Refuge Plus Option Planter

Introduction to Crop Production is written for those with a desire to revisit our agricultural roots — but in the context of the very modern industry into which agriculture has grown. The objective of this book is to provide a brief overview of the processes, principles, and practices associated with production of the major crops grown in the United States. The major areas covered in this book are:

- Soil Management — Crop production starts with soil management. Feed the soil, the soil feeds the plants, and your plants feed you. We will cover the main nutrients that plants need to grow, along with ways to add these nutrients to the soil with the goal to grow a productive crop.
- Tillage — Tillage is a key component to crop production. There are a variety of ways to till the soil and many ways in which tillage helps to keep a good growing environment for the crop. We will discuss the tools that modern farmers use for tillage and explain how tillage impacts plant growth.
- Planting — Planting is a major element in crop production, because without placing the seed correctly in the soil it would be impossible to produce a crop. Major aspects covered in this book regarding planting are the types of planting equipment a crop producer may use along with proper planting techniques to ensure a profitable stand of plants.

Continued on next page

OUO1023,00040C2 -19-22OCT15-4/5

102615
PN=8

- Chemical Application Management — Chemicals are used in a variety of ways to help the crop thrive. At times, crops may need additional nutrients added to the soil, which can be accomplished through the use of chemicals. Chemicals many times are used to correct problems in the farm field along with prevent those problems from starting. One may have heard the old saying "an ounce of prevention is worth a pound of cure." The use of chemicals can put money in the crop producer's pocket due to increased crop yields. Though there are multiple benefits to using chemicals in agriculture, one must be cautious — there are also hazards.
- Combine Harvesting — Harvesting the crop brings all the hard work that was done in the growing season to a close. It is essential to have a skilled operator and a well-adjusted combine during harvesting season. Profits depend on multiple uncontrollable factors, but one of the factors a crop producer can control is the way in which the crop is harvested. One must do everything possible to handle the crop correctly and harvest without error. The combine harvesting chapter explains key components and operations of the combine, and advises on where profits can be maximized.
- Hay and Forage Harvesting — There are multiple varieties of hay and forage that can be grown to feed livestock. Harvesting hay and forage crops requires that the farmer preserves as much nutritional value in the crop as possible, with the lowest investment of labor and money. The types of equipment used for hay and forage harvesting along with how crops are handled during harvest are explained in this chapter.
- Precision Farming — Over the history of farming it has been discovered that profits can be maximized when cropland is managed on a site-specific basis. This chapter will discuss managing each crop production input — fertilizer, limestone, herbicide, insecticide, seed, etc. — on a site-specific basis to reduce waste, increase profits, and improve the quality of the environment. By implementing precision-farming concepts in a farming operation, it allows farmers to become better stewards of their farmland.
- Machinery Management — The reasons why farmers decide on a type of farm machinery to accomplish work depends on many factors. A good farm manager will carefully plan and use the best machinery alternative for his or her operation. At times, a custom operator may be hired to maximize profits. The machinery management chapter will discuss how to decide on the best machinery alternative as well as discuss ways to estimate costs incurred with machinery.

Crop production is a rewarding industry to understand and support; this publication provides a basis for that growth to begin.

OUO1023,00040C2 -19-22OCT15-5/5

Soil Management

Introduction

DXP01865 —UN—08DEC10

Higher-yielding crops require fertile soil that provides a suitable environment for seed germination and improves root growth, with minimal competition from weeds and with adequate but not excessive moisture. However, today's crop producers are faced with many challenges: rising fuel costs and government regulations force some producers to minimize their tillage passes, while other producers have tough residue, requiring additional sizing and mixing to adequately break down. Some producers who adopted conservation tillage are finding that compaction limits them from achieving optimum yields.

This chapter will discuss the role of soil in providing nutrition to the crop for maximum yields, along with the role of management, including aspects of tillage in minimizing compaction, managing residue, and providing an ideal seedbed.

MM16633,000259E -19-09DEC10-1/1

Soil Fertility

Nutrients	Category	Comments
Carbon, Hydrogen, Oxygen	—	Required in greatest quantities and are typically supplied by the soil and air; certain factors such as compaction can limit availability.
Nitrogen, Phosphorous, Potassium	Primary macronutrients	Commonly referred to as N-P-K. Required in large amounts for ideal crop production; without adequate amounts, crop production is limited
Calcium, Magnesium, Sulfur	Secondary macronutrients	Required in large amounts; most soils are able to supply adequate amounts for crop growth
Boron, Chlorine, Copper, Iron, Manganese, Molybdenum, Zinc	Micronutrients	Required in the least amount; typically added when soils are deficient and/or crops are particularly sensitive

Table 1 — Nutrients Needed by Plants

Table 1 — Nutrients Needed by Plants

Soil is a medium for plant growth and is a mixture of mineral particles, organic matter, water, and air. An ideal soil for plant growth has the following characteristics:

- At least 60 inches of soil material favorable for root growth
- Approximately 45% mineral matter, 5% organic matter, and 50% pore space, filled with equal volumes of water and air
- A loamy texture
- An aggregated structure permitting easy water entry, deep root penetration, sufficient air movement, and internal drainage
- Micro-organisms and small life-forms such as earthworms to decompose organic matter
- An adequate and balanced supply of plant nutrients (Table 1)
- Correct soil acidity, or pH, which influences the availability of plant nutrients and the performance of some pesticides

Plants, like animals, need food to help them grow and develop. At least 16 elements are needed for a plant to grow normally. If any one of the essential elements is missing, the plant will die. If an element is in short supply, the plant will be undernourished and will not grow properly. Well-fed plants provide better nutrition for animals or people who consume the crop.

Most of the elements needed for plant growth are supplied by the soil, air, or water. For example, carbon, hydrogen, and oxygen are required in large quantities but available in the air and water. The soil usually can provide the balance of the nutrients, but some, like nitrogen, phosphorus, and potassium are required in relatively large amounts that the soil may not adequately supply.

Fertilizers are materials which supply one or more of the essential plant food elements. Many materials have been used as fertilizers but the most common natural material that is used as fertilizer is animal manure.

Legumes such as alfalfa, soybeans, clover and peanuts also supply nitrogen, an essential nutrient. Soil bacteria form nodules on legume roots. Through the process of fixation, well-nodulated legumes convert atmospheric nitrogen to ammonium nitrogen which can be utilized by the plant.

MM16633,000259F -19-08OCT10-1/1

Effect of Soil pH

Most soils in humid regions are acid. The degree of acidity is expressed as soil pH. The pH scale has 14 units. A pH of 7.0 is neutral. Lower pH indicates acid soil; higher pH indicates alkaline soil.

The pH scale is logarithmic. That means a pH of 5.0 is 10 times as acid as a pH of 6.0. Similarly, pH of 9.0 is 10 times as alkaline as pH 8.0. Soil pH has a great effect on the availability of each of the soil-derived nutrients discussed earlier, and can either restrict availability or even cause them to be toxic. Generally, all elements are available at optimum levels in a neutral soil. This is why most agricultural crops grow best in soil with pH near 7.0.

However, there are exceptions. Crops which grow better with lower pH include blueberries and cranberries. Alfalfa and sweet clover thrive at pH levels above 7.0. Because plant nutrient availability is closely associated with soil acidity, soil pH can also affect plant response to fertilizer. If soil is too acid or too alkaline, the pH can be changed by adding the proper material.

Correcting Soil pH

Lime Corrects Soil Acidity — The standard way to correct soil acidity is to add crushed limestone to the soil. Lime supplies some plant food elements, but is not generally thought of as a fertilizer. Lime raises the pH of acid soil so plants can extract food elements more easily. The relationship between lime and plant food availability has led to a commonly used expression, "Lime and fertilizer work as a team."

Gypsum and Sulfur Correct Soil Alkalinity — While acid soil is a common problem in humid areas, alkaline soil Is a concern in arid areas. To lower soil pH, materials such as gypsum and sulfur are applied.

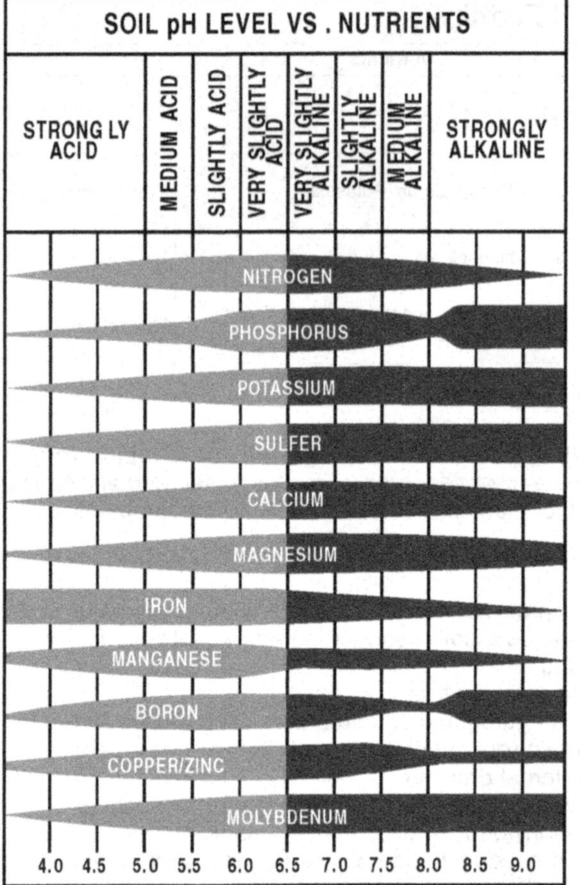

Soil pH affects nutrient availability

MM16633,00025DA -19-27OCT10-1/1

Primary Macronutrients

Three nutrients — nitrogen, phosphorus and potassium — are often applied in large quantities to soils to maintain and improve crop growth. These three nutrients are present in the plant in larger quantities than other nutrients and are called the primary macronutrients.

Nitrogen

Crops use large amounts of nitrogen. In fact, more nitrogen is used as plant food than any other two mineral elements combined.

Nitrogen is essential for cell division and plant growth. It is an element in proteins, amino-acids, sugars. and chlorophyll, and is involved in plants' synthesis of sugars and starches which do not contain nitrogen.

Nitrogen promotes rapid leaf and stem growth and gives plants succulence and a dark green color. If nitrogen is limited, plants are small and usually have poor color; corn ears may have unfilled tips.

Atmospheric Nitrogen

Approximately 80% of the atmosphere is nitrogen gas. There are 35,000 tons of nitrogen in the air above every acre (78,400 metric tons above each hectare). Nitrogen is constantly being removed from and returned to the atmosphere in a group of processes collectively known as the Nitrogen Cycle.

Obviously, the supply of nitrogen in the atmosphere is adequate for any crop needs we can foresee. However, atmospheric nitrogen is a relatively inert gas that most plants cannot use directly. These plants must rely on

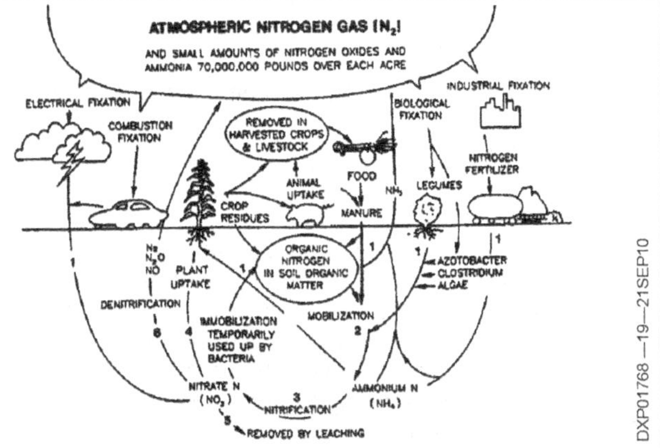

The Nitrogen Cycle

nitrogen that has been "fixed" by a biological or industrial process. Fixed nitrogen is combined with other elements to form compounds which the plant can use. Legumes are plants that can use atmospheric nitrogen that has been fixed by soil bacteria which grow symbiotically on the legume roots. Research is underway to transfer its nitrogen-fixing capability to corn and other non-legumes. However, the amount of nitrogen that can be fixed biologically is currently not sufficient to provide for maximum growth of most crops. Consequently, most of the nitrogen used by crops is manufactured and applied as commercial fertilizer.

Plants can use two forms of nitrogen.

- Ammonium-nitrogen (combined with hydrogen)
- Nitrate-nitrogen (combined with oxygen)

Continued on next page

MM16633,00025DB -19-19NOV10-1/5

PN=13

Forms of Soil Nitrogen

The three principal forms of soil nitrogen are:

- Organic nitrogen
- Ammonium nitrogen
- Nitrate nitrogen

Organic nitrogen — The largest portion of nitrogen in most soils is organic nitrogen. The top 7 inches (18 cm) of mineral soils contain 2,000 to 20,000 pounds of nitrogen per acre (2,240 to 22,420 kilograms per hectare). Peat soils, which are up to 95% organic matter, may hold up to 50,000 pounds of nitrogen per acre (56,050 kg/ha). However, only 1% to 2% of this nitrogen is released for crop use during a normal growing season. This amounts to 30 to 40 pounds per acre (34 to 45 kg/ha) for typical Corn Belt soil. The remaining crop needs are met by application of nitrogen fertilizers.

Organic nitrogen exists in many forms. Much of is undergoing a constant process of change from one form to another. The action of soil microorganisms can release ammonia. The ammonia is then held in the soil until it is taken up by plants or lost to the atmosphere.

Research has shown that plants respond identically to nitrogen whether the original source is organic or inorganic. However, nitrogen from organic matter has several important advantages over other nitrogen fertilizers:

- It is released slowly

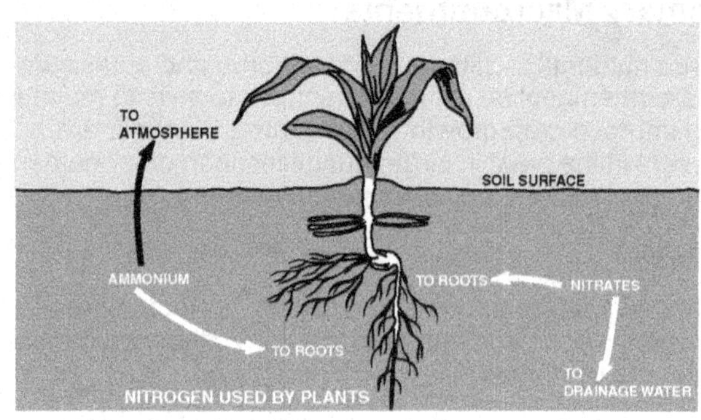

Form of nitrogen used by plants

- It is not lost by leaching
- It is released continuously

Ammonium nitrogen — This form of nitrogen may either be applied to the soil or be produced by soil bacteria. Soils will absorb and hold ammonium nitrogen for only a short time. The holding ability depends on soil type and ranges up to 2,000 pounds per acre (2,240 kg/ha) for clay. Ammonia is held weakly on soil particles and can be released by soil micro-organisms. Then the ammonium nitrogen is reabsorbed by soil, used by plants, changed to nitrates, or is lost to the air. The activity of soil bacteria speeds up as the temperature rises above 50°F (10°C). Unless there is growing crop to use the displaced or released nitrogen, much is lost.

Continued on next page

MM16633,00025DB -19-19NOV10-2/5

Nitrate nitrogen is supplied in fertilizer or produced in the soil by transformation from the ammonia or ammonium forms. The transformation from the ammonium form, called nitrification, occurs in two steps:

- Ammonium nitrogen changes to nitrites
- Nitrites change to nitrates

Nitrates are held in the soil solution. Thus, they are more readily available to the plants than nitrogen in the ammonium form. However, nitrates may also be lost easily in drainage water.

Nitrates can also be lost if the process of denitrification occurs. The N_2O and N_2 gases formed in the volatilization process can escape through the soil surface.

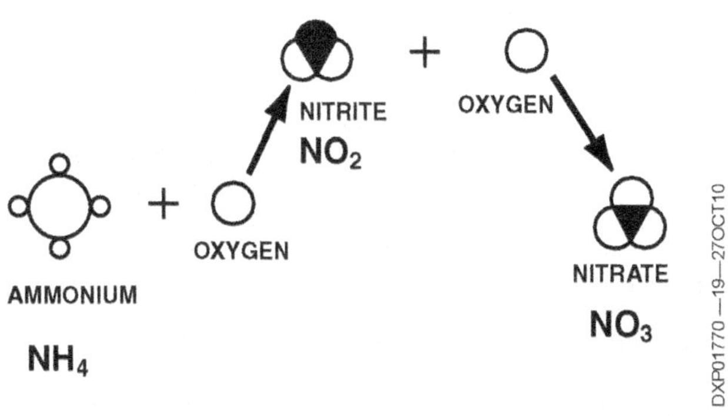

Nitrification process

Major Nitrogen Fertilizer Forms

Fertilizer	Abbreviation	Description	% N by weight	Application method
Anhydrous ammonia	NH_3	Pressured liquid that quickly converts into gas. Once injected into the soil, it quickly absorbs water and converts to ammonium, the most stable form of nitrogen	82%	Injection
Urea	$CO(NH_2)_2$	Dry material containing nitrogen. Urease converts urea to ammonia and then to ammonium	45–46%	Surface application with tillage incorporation (tillage incorporation is not necessary if rain is received shortly after application)
Nitrogen solutions	UAN	Liquid solution of urea and ammonium nitrate	28–32%	Inject, spray, or dribble on soil surface
Ammonium nitrate	NH_4NO_4	Dry prill or granular form. Nitrate is immediately available to plants	34%	Surface application

Table 2 — Major Nitrogen Fertilizer Forms

Major nitrogen fertilizer forms

Conserving Nitrogen

Getting maximum results from nitrogen requires knowledge of factors that cause significant losses. They are:

- Leaching
- Denitrification
- Volatilization

Leaching — Leaching occurs when water moves through soil and carries nitrate nitrogen below the root zone. In the ammonium form, nitrogen clings to clay particles in the soil. Soil micro-organisms convert ammonium nitrogen to nitrate nitrogen, a process known as nitrification. Nitrate nitrogen is available for crop growth, but is also subject to leaching.

Leaching is a problem on sandy soils since they are more porous than clay soils. To minimize losses, a farmer can make a split application. Part of the nitrogen is applied just before or at planting. Then the rest is "sidedressed" by applying it on either side of the crop row during the growing season.

Taking steps to stop leaching is important for two reasons. Besides conserving nitrogen for crop growth, it also keeps nitrates from contaminating ground water.

Several products are available to slow nitrification. They kill soil microorganisms involved in the conversion of ammonium nitrogen to nitrate nitrogen. In university research, the nitrification inhibitors have given mixed results. However, they are worth considering when conditions favor nitrification, particularly soils warmer than 50°F (10°C).

Continued on next page

MM16633,00025DB -19-19NOV10-3/5

Denitrification — Denitrification is the loss of nitrogen from soil to the atmosphere. It occurs when soil becomes waterlogged and the oxygen supply is depleted. Soil micro-organisms utilize oxygen. When it is lacking, they feed on nitrate nitrogen and change it to a gas. To control denitrification, fields with poor drainage can be tiled. Irrigated fields need to be leveled to keep water from collecting in low spots. The amount of irrigation water applied must also be properly controlled.

Volatilization — Volatilization occurs when ammonium nitrogen or urea are applied to the soil surface and converted to ammonia, which is a gas. Ammonium nitrogen is subject to volatilization if soil pH is above 7.0. With urea, the problem exists regardless of soil pH when it remains on the soil surface. The answer is to incorporate either form of nitrogen into the soil.

Phosphorus

Crop	Yield	Phosphorus (lb.)
Alfalfa	6 tons	70
Apples	600 bu.	15
Barley	100 bu.	40
Corn	150 bu.	50
Cotton	3,200 lb. of lint and seed	30
Oats	100 bu.	25
Oranges	800 boxes	30
Peaches	600 bu.	10
Potatoes	400 bu.	35
Soybeans	50 bu.	40
Sugar Beets	30 tons	50
Timothy	3 tons	30
Tobacco	2,800 lb. of stems and leaves	25
Tomatoes	30 tons	55
Wheat	60 bu.	35

Table 3 — Pounds of Phosphorus in Crops

Phosphorus, as rock phosphate, was one of the earliest chemical fertilizers. Though the use of nitrogen is now greater, phosphorus is still an important fertilizer (Table 3).

Phosphorus stimulates early root growth and blooming, and hastens crop maturity. It contributes to strong plant structures, improves seed quality, and increases resistance and winter hardiness in some plants.

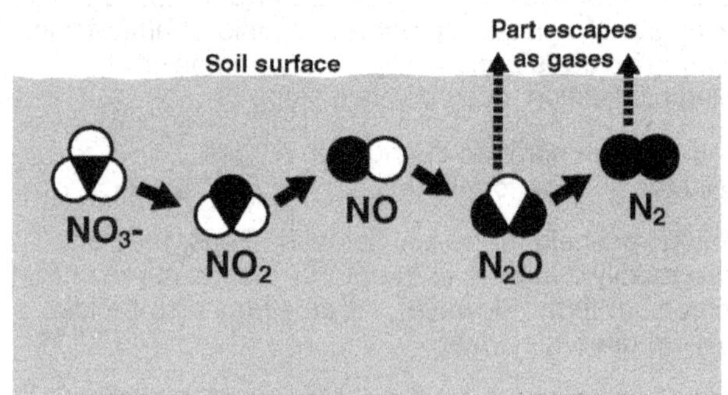

Denitrification process

Phosphorus-deficient plants are unable to use other nutrients properly. A common symptom of a phosphorus shortage is a buildup at nitrogen in the lower leaves, which then take on a purplish color. When a plant develops a phosphorus deficiency, it usually is stunted and does not recover completely. Animals fed phosphorus-deficient plants may also have symptoms of phosphorus deficiency, such as low vitality, poor reproductive efficiency, and weak bones.

Plants obtain phosphorus from the soil solution. The rate of phosphorus consumption depends on the stage of growth. It has long been known that an application of phosphorus fertilizer at planting time helps get a crop off to a rapid start.

Phosphorus Fertilizers

As a pure element, phosphorus is a material which bursts into flame when exposed to air. So, obviously, the pure element cannot be used for fertilizing. It must be combined with other elements to make it safe to handle and useful to crops.

In the fertilizer trade, the phosphorus content of fertilizers is expressed in terms of percent phosphate (P_2O_5). A list of the most common phosphorus-bearing fertilizers are shown in Table 4. The basic material is rock phosphate which is mined at locations around the world. Various materials are then combined to make fertilizers. Some phosphorus fertilizers, such as MAP and DAP supply nitrogen to the crop as well as phosphorus.

Fertilizer	N	P_2O_5	K_2O
MAP (monoammonium phosphate)	11	48/52	—
DAP (diammonioum phosphate)	18	46	—
Ammonium polyphosphate	10	34	—
Triple superphoshate	0	45	—
Potassium chloride (muriate of potash)	—	—	60–62
Potassium sulphate	—	—	50
Potassium magnesium sulphate	—	—	22

Table 4 — Commercial Fertilizers Containing Phosphorus and Potassium

Table 4 — Commercial Fertilizers Containing Phosphorus and Potassium

Continued on next page

MM16633,00025DB -19-19NOV10-4/5

Water Solubility

Material Analysis	Water Solubility
Ammonium Phosphate (11-48-0)	89%
Ammonium Phosphate (16-20-0)	86%
20% Superphosphate (0-20-0)	78%
Triple Superphosphate (0-47-0)	84%
Dicalcium Phosphate (0-40-0)	5%
Calcium Phosphate (0-62-0)	4%

Table 5 — Water-Solubility of Phosphorus Materials

Table 5 — Water-Solubility of Phosphorus Materials

Water solubility of phosphorus determines its availability to plants. Solubility of phosphorus-bearing materials can vary widely (Table 5) and this factor has a marked effect on plant response. Too much phosphorus in the soil (over-fertilization) can cause iron and zinc deficiencies in many crop plants.

Loss of Water Solubility

Phosphates tend to change into insoluble forms that are unavailable to plants. The first crop seldom is able to take up more than 30% of a phosphorus application. Each year, the percentage of recovery decreases. Consequently, applications of phosphate fertilizers every two to three years are needed to help ensure adequate phosphorus for each crop.

Potassium

Potassium is absorbed by plants in greater quantities than any other nutrient except nitrogen. It is found in cell fluids throughout the plant, but is not built into the plant structure. Symptoms of potassium deficiency include curling, yellowing, "scorching" or bronzing of leaf margins and tips; stems are weak; roots and tubers (such as potatoes) are undeveloped; and plants are often stunted.

Potassium is sometimes described as a "chemical policeman" which keeps things moving. The functions of potassium include:

- Increases root and stem growth
- Improves drought resistance
- Improves product quality
- Helps retard diseases
- Builds cellulose, adds stem or stalk strength, and helps prevent lodging
- Aids formation and movement of starches and sugars
- Enhances protein production
- Aids in photosynthesis
- Reduces water loss and wilting

Potassium Fertilizers

Pure potassium is a light gray metal that burns when exposed to air. Like nitrogen and phosphorus, potassium must be combined with other elements before plants can use it.

In the fertilizer trade, the potassium content of fertilizer is expressed in terms of potash (K_2O).

Many potassium-bearing materials may be used as fertilizers. A list of common potassium-fertilizer materials is included in Table 4.

Potassium Availability

Most U.S. soils contain large quantities of potash — up to 40,000 pounds per acre (44,800 kg/ha) in the plow layer. However, most of it is not in an available form for plant use. Most crops place a fairly heavy drain on the supply of potash in the soil. When crops are produced successively on the same land, the soil potash is gradually depleted and crop yields decline unless additional amounts are made available. Potassium shortage appears in corn as yellow and brown leaf margins, appearing first on the lower leaves before moving to upper leaves.

MM16633,00025DB -19-19NOV10-5/5

Secondary Macronutrients

Secondary nutrients are essential to plant growth, but are required in smaller amounts than the three primary nutrients. The amount of each secondary nutrient needed depends on:

- Plant species
- Soil conditions
- Climate
- Availability

Secondary macronutrients include:

Calcium

Calcium stimulates root, stem, and leaf growth, and improves general vigor and disease resistance. It increases the uptake of other essential nutrients and is needed to ensure crop seed production and maturity.

Calcium improves soil structure by causing aggregate-forming granules of soil. Granular soil provides a good root environment by allowing easy penetration of air and water.

Calcium is also used to correct soil acidity. This need becomes more evident as soils become older. Crops have absorbed a great deal of calcium from the soil and large amounts have been removed by leaching due to rainfall. The tremendous increase in the use of acid-forming nitrogen fertilizers has also intensified the need for calcium.

NOTE: Excessive calcium in the soil may cause deficiencies of boron, iron, magnesium, manganese, potassium, and zinc.

Ground and burned limestone and gypsum are the most widely used calcium sources.

Magnesium

Magnesium is essential for production of chlorophyll. It aids in formation of many plant compounds such as sugars, starches, oils, and fats. It also has a role in movement of phosphorus and other plant foods to various parts of a plant. The most common source of magnesium is dolomitic limestone, which contains both magnesium and calcium, although other magnesium compounds are also used.

Sulfur

Sulfur is found in all living plant cells as part of the protoplasm. Plant seeds have a higher concentration of sulfur than other plant parts. Sulfur aids in formation of proteins, encourages vigorous growth, and helps plants withstand cold temperatures.

At one time, coal and oil were used extensively to heat homes and supply power for factories and utilities. When coal and oil are burned, sulfur is released into the atmosphere. It is then returned to the earth by rain or snow.

Sulfur deficiencies have become common as cleaner burning fuels have replaced coal and oil. In addition, modern high-yielding crop varieties and hybrids have depleted soil of sulfur.

Commercial fertilizers are used to correct sulfur deficiencies. Sources include ammonium sulfate and superphosphate. Some sulfur is also supplied by decomposing plant matter and irrigation water.

MM16633,00025DC -19-29SEP10-1/1

Micronutrients

Micronutrient	Soil Conditions	Crops
Boron	Sands, overlimed acid soil, organic soil	Cotton, tomatoes, citrus, sweet potatoes, leafy vegetables, tree fruits, legumes with heavy lime.
Copper	Sands, organic soils, high pH	Small grains, vegetables, tree fruits.
Iron	High pH soils, high phosphate	Blueberries, corn, sorghum, soybeans, ornamentals.
Manganese	Sands, overlimed soils, organic soils	Soybeans, small grains, tree fruits, cotton, sweet potatoes, leafy vegetables.
Zinc	Sands, high pH soils, high phosphate	Soybeans, corn, citrus, rice, sorghum, pecans, tree fruits, some vegetables.
Molybdenum	Highly weathered acid soils	Legumes, citrus, cauliflower, cabbage, and similar plants.

Table 6 — Micronutrient Deficiencies Affect Many Crops

Table 6 — Micronutrient Deficiencies Affect Many Crops

Micronutrients, also called trace elements, are applied in very small amounts. Deficiencies are found in all areas of the U.S. and affect many crops. Most micronutrient-deficient plants exhibit two symptoms:

- Stunted growth
- Discoloration of foliage

Diagnosis of particular deficiencies requires experience as well as knowledge of the symptoms. Tissue and soil tests can confirm visual evaluation. These tests can also be used to detect deficiencies before the symptoms become visible. Corrective action can then be taken before serious growth impairment results.

MM16633,00025DD -19-08OCT10-1/1

102615
PN=18

Fertilizer Analysis

Fertilizers may contain more than one essential element and may be blended to meet the specific needs of the field or crop. The amount of nitrogen (N), phosphorus (P) and potassium (K) represented in a fertilizer are referred to as the analysis, and are expressed as the percent of the total product weight of N-P-K. For example, (see table), 100 pounds of the phosphorus fertilizer DAP contains 46 pounds (46%) of phosphorus, but also contains 18 pounds of nitrogen. The analysis of DAP is expressed as 18-46-0, as it contains 18, 46, and 0 pounds of nitrogen, phosphorus, and potassium, respectively.

To manage fertility under reduced tillage systems:

- Build the soil phosphorus, potassium, and lime levels to suggested goals throughout the plow layer before starting a conservation tillage program
- Monitor soil test levels every two to three years
- If nutrient distribution becomes a problem, periodically include a tillage practice that will thoroughly mix the upper six to eight inches of soil
- Use an injected nitrogen source if possible
- Consider using a starter fertilizer for corn at planting time, especially if nitrogen is preplant-injected or side-dressed
- If nitrogen must be broadcast on the soil surface without incorporation, do it when the probability is high of receiving rain in the next few days

MM16633,00025DE -19-29SEP10-1/1

Soil Erosion

What is soil erosion?

Soil erosion is the detachment and movement of soil or rock caused by water, wind, ice, or gravity. In agriculture, the two major causes of soil erosion are water and wind.

What are the effects of soil erosion?

Erosion removes a significant amount of topsoil and leaves the subsoil, which is not as productive as topsoil. This is a serious issue in agriculture because topsoil, which takes at least 150 to 300 years to form, contains the highest concentrations of organic matter and plant nutrients. Organic matter contributes to soil conditions favorable for seed germination, emergence, root growth, and water intake and storage. Typically, five to ten inches of topsoil plus an adequate subsoil layer are needed for efficient growth of ordinary field crops.

In addition to the loss of nutrients, soil erosion may:

- Contaminate surface waters with fertilizers and pesticides
- Change the physical makeup of the field. For example, topsoil has a very different physical makeup than subsoil. When topsoil is eroded from the top of a hill and deposited at the bottom of the hill, the nature of the hilltop with shallow topsoil is different than the soil at the bottom of the hill. This causes plants at the hilltop to require more nutrients than those plants at the bottom of the hill.
- Impact yields
- Require repairing or cleaning up, which is an additional cost to the producer

How does water erode soil?

Water erodes soil in four ways:

1. Splash erosion occurs when droplets of water land directly on soil or on thin films of water covering the soil, dislodging soil particles and making it easier for soil to be washed away. On bare soil, it is estimated that as much as 100 tons of soil per acre are dislodged by heavy rains.
2. Sheet erosion is caused when a sheet of water flows across a field, moving a thin layer of soil. Raindrops cause soil particles to become detached (splash stage) and the flowing water moves the soil down hill.
3. Rill erosion is caused when running water creates a small channel in the field, dislodging soil and moving it to lower part of the field. As soil detachment continues or water flow increases, rills will become wider and deeper.
4. Gully erosion is an advanced stage of rill erosion, just as rills are often the result of sheet erosion. The channels are much larger and cannot always be fixed by tillage.

How do I reduce the effects of water erosion?

Tillage operations can repair damage caused by water erosion. It also can help control sheet erosion as a rough surface profile, along with crop residue, makes it difficult for water to move soil. Minimum-till operations will help reduce splash erosion as more residue is left on the soil surface, protecting the soil from hard rains.

A gully that can be removed by tillage is classified as an "ephemeral gully." An ephemeral gully may recur. Creating dams or terraces, or using other conservation measures, can help control or stop a gully.

How can I reduce or control wind erosion?

- Primary or secondary tillage operations should be tailored to produce a ridged or rough, cloddy surface to absorb or deflect part of the wind energy. The depressions will trap drifting soil particles.
- Residue management which keeps a sufficient amount of residue on the surface
- Planting methods such as rotation, row width, and row direction
- Harvesting practices such as using a spreader on the combine to evenly distribute residue
- Building windbreaks

How do I determine residue levels?

Two ways to determine residue levels are: the line-transect method (to measure existing levels) and the prediction method.

- The line-transect method measures the amount of soil surface covered with residue at measured points along a line. It is the most accurate and recommended method of determining residue levels.
- The prediction method estimates the amount of residue that will remain on the soil surface after one or more field operations. It is computed by multiplying the initial residue level times factors which estimate the amount of residue which would remain after each particular tillage or planting operation.

MM16633,00025A0 -19-27OCT10-1/1

Compaction

What is compaction?

Compaction occurs when pressure from farm animals or equipment increases soil density. The solid portion of soil is made of particles of sand, silt, clay, and organic matter. The organic matter bonds particles together, forming aggregates. Large pore spaces surrounding the aggregate are filled with air or water. A good root bed contains 50% solid material, 25% water, and 25% air. However, this pore space is reduced in compacted soil.

While some "compaction" can be beneficial for achieving seed-to-soil contact, most compaction can be detrimental to crop growth and yield. Field traffic, not tillage, is the major cause of compaction, especially from operations that involve heavy loads on damp or wet soil. Water acts as a lubricant, letting soil particles compress easier. Soil that contains a lot of clay or about equal proportions of all particle sizes is more prone to compaction.

The amount of compaction depends on:

- Soil structure and type of soil
- Moisture level
- Use of tires or tracks on tractors
- Inflation pressure of tires
- Weight of equipment, farm animals, etc.

What types of compaction are there?

- Surface crusting is the formation of a thin, hard layer on the soil surface that "seals" the soil surface from water and air infiltration, greatly inhibiting crop emergence. This crusted layer can be one-half to an inch thick. It is caused by the impact of heavy rain on weak soil aggregates. Soils with high organic matter, high biological activity, or high sand content are less likely to form crusts.
- Surface compaction exists in the top 8 to 12 inches of soil. It can be loosened with normal tillage, root growth, freeze/thaw cycle, wet/dry cycle, and biological activity. The amount of compaction is based on the moisture in the soil upon compaction and the ground-contact pressure of equipment or animals (measured in pounds per square inch, or psi).
- Subsoil or deep compaction is beneath the tillage depth. Ground-contact pressure and the total weight on the tire (the axle load) significantly affect the amount of subsoil compaction. Deep compaction is difficult to eliminate and may permanently change the soil structure. Prevention is important.
- Hard pan or tillage pan is subsoil compaction only a few inches thick and lies directly beneath the normal tillage depth. It develops when the depth of tillage is the same from year to year and too shallow to break deep compaction, or when the rear wheel of the tractor rides in the moldboard plow furrow.
- Sidewall compaction is the slicking and compacting of the sidewalls of the seed trench by the planting

Pressure on soil

Crushes soil aggregates

Reduces pore space

Slows exchange of oxygen and carbon dioxide

Limits water and nutrient movement

Compresses air pockets

Lowers temperature of soil

Increases bulk density

Increases soil strength

Slows root penetration

Increases power requirements for tillage

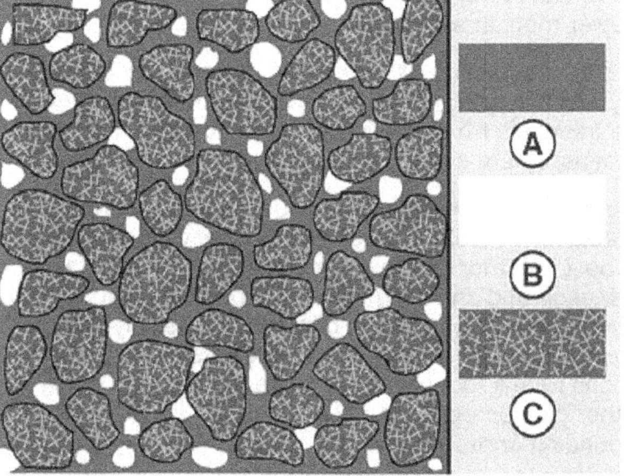

Development of soil compaction

Soil Composition

A— Water
B— Air
C— Aggregate

equipment. It is most prevalent in heavy, wet, or muddy planting conditions. With slick, hard sidewalls along the seed trench, the roots will follow the seed trench instead of achieving proper root formation.

What soil structures are more susceptible to compaction?

- Loamy-sand and sandy-loam soils are more resistant to soil compaction because they are moderately coarse-textured sands with enough organic matter to provide structure.
- Loam, silt-loam, and silty-clay-loam soils are relatively resistant to compaction. The more clay content, the more susceptible to compaction the soil is (especially if organic matter content is low).
- Soils with more than 45% clay and clay-like clay loam are quite susceptible to compaction because they have a large portion of pore space, although the pores are very small and remain filled with water.

Continued on next page

MM16633,00025A1 -19-08DEC10-1/3

How do I know if I have compaction?

Look for the following signs in your plants:

- Slow crop emergence
- Uneven stands
- Off-color or purple leaf discoloration
- Uneven plant heights or signs of plant stress
- Early yellowing that might indicate nitrogen loss in crops
- Shallow, constricted roots
- Malformed roots

In addition, look for the following signs in your soil:

- Reduced water infiltration
- Standing water
- More surface erosion
- Fewer roots in soil profile

Signs of compaction usually do not cover a whole field. They begin in wetter areas or where the heaviest vehicles have traveled. An exception is the sandy soils of the southeast where signs may first appear in droughty soils.

Keep in mind that what appears to be signs of compaction can actually be nutrient, disease, or herbicide stresses. The best indicator comes from digging plants midway in the season and checking roots. By this time, signs of early disease and herbicide damage are usually past. Digging should be deeper than the tillage zone. If soil sections reveal root problems, poke the soil with a knife or other pointed object to check if the soil is denser than surrounding areas with healthy roots.

Pancake root mass

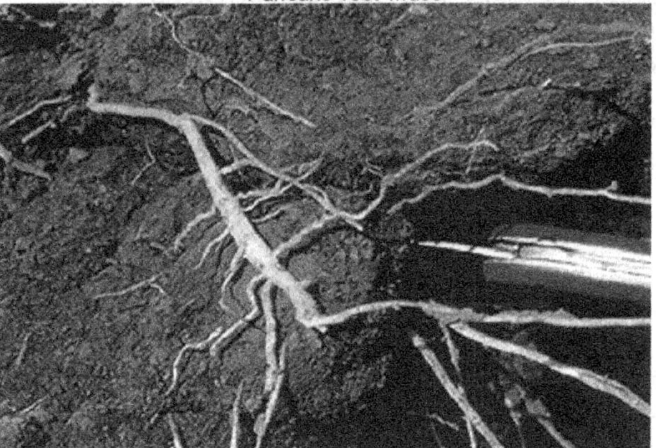

Flattened root caused by compaction

Continued on next page

MM16633,00025A1 -19-08DEC10-2/3

How do I prevent compaction?

- Eliminate traffic on wet soils. To determine if your soil is too wet, form a ball of soil and throw it in the air. If the ball breaks apart, the soil is dry enough to drive on. Another way to test soil is to ribbon some soil between your thumb and index finger. If the ribbon breaks before it reaches five inches in length, the soil is dry enough to drive on.
- Control traffic patterns in the field. Up to 80% of compaction occurs on the first pass. Keep heavy equipment (e.g., loaded grain carts, trucks) to the smallest area of the field as possible, and use the same tracks with each pass. Use the John Deere Ag Management System (AMS) to map passes and stay on track. In addition, infrequent operations with heavy machines can be performed during dry periods if necessary.
- Configure equipment to reduce compaction, using consistent tire width and spacing on all equipment. Reducing axle loads, and reducing tire pressure to increase the size of the tire print can reduce compaction. This is especially helpful on firm soils. For example, you can use dual tires, tracks, or wide tires to increase the area that contacts the soil. In addition, correct tire pressure will greatly improve tractive efficiency.

How do I correctly set depth on my tillage tool?

- Determine compaction depth using a soil probe or knife.
- Run the tool in the ground at set depth (make sure that the single-point plate is touching the plunger if your implement has this feature).
- Run at proper operating speed (see operator's manual).
- Stop with the tool in the ground and put the tractor in park.
- Smooth out soil around the shank to ground level—it is important that depth be measured from ground level, not the soil blow-out level.

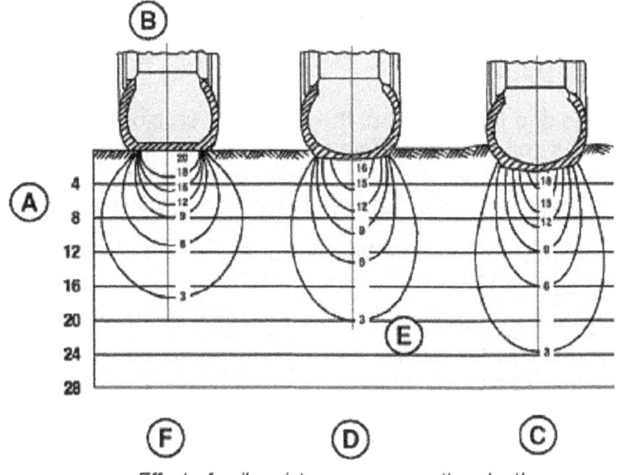

Effect of soil moisture on compaction depth

A—Distance Below Soil Surface (Inches)
B—Tire Size 11-28 Load: 1650 lbs Inflation Pressure 12 PSI
C—Wet Soil
D—Soil with Normal Density and Water Content
E—PSI
F—Hard Dry Soil

- Use the soil probe behind the shank to determine how deep the shank is going by lightly pushing into the slit that the ripper made until you feel the bottom.
- Adjust your depth control to be 1 to 2 inches below the hard pan.
- Perform another pass and check depth to verify proper settings.

MM16633,00025A1 -19-08DEC10-3/3

Residue Management

Crop Residue

Crop residue helps control the effects of erosion and maintains soil structure. The amount of crop residue on a field plays an important part in determining what crops are planted next and what field operations are performed.

Leaving crop residue on the soil surface can provide the following benefits:

- Reduced soil erosion
- Improved water quality
- Conserved soil moisture
- Increased organic matter in soil
- Reduced air pollution
- Improved soil tilth
- Decreased soil compaction
- Reduced machinery wear
- Reduced fuel usage

Residue management includes all field operations that affect residue amount, orientation and distribution.

Managing Crop Residue

Crop residue can be managed by:

- Crop selection
- Crop rotation
- Residue removal
- Tillage practices
- Combine settings

Crop Rotation and Residue Levels

Grain Type	Residue Index (pounds per bushel)
Winter wheat	80–110
Spring wheat	70–100
Corn	60
Grain sorghum	55–60
Oats	40–60
Soybeans	45–75
Sunflower	1.5
Barley	50–65
Millet	80

Table 7 — Residual Indexes for Estimating Agricultural Residue Protection

Table 7 — Residual Indexes for Estimating Agricultural Residue Protection

Residue can be a major benefit to the cropping system; however, it also can produce many problems if not managed properly. Proper crop rotation can help manage residue levels. By rotating high-residue crops with low-residue crops, one can manage residue, control weeds and crop diseases, and improve soil fertility. An example of a crop rotation is cereal, oilseed, pulse, cereal. The specific crops grown in a rotation will vary by region.

Residue Management Options During Harvest

Crop residue can be managed by harvest practices. The harvest operation is the first and best time to manage residue by sizing and evenly distributing it. Combine configuration will determine the size and condition of residue left on the field.

Keep the following factors in mind:

- Cutting height
- Use of harvest tools to evenly distribute residue, such as:
  - Shredders
  - Stalk choppers
  - Stalk rolls (fluted or knife)
  - Straw spreaders and choppers
  - Chaff spreader

MM16633,00025A2 -19-09DEC10-1/1

Introduction

Tillage has been defined as those mechanical, soil-stirring actions performed for the purpose of nurturing crops. The goal of proper tillage is to provide a suitable environment for seed germination, root growth, weed control, soil-erosion control, and moisture control.

Tillage requires well over half of the engine power expended on American farms, and it has been estimated that more than 250 billion tons of soil are tilled each year in this country. Many of the implements used, and much of the need for all this soil movement, have long been taken for granted. There have been few revolutionary developments in tillage equipment. Most changes have been the result of modification, improvement, and evolution of earlier equipment designs and ideas. Despite few developments, there have been more changes in tillage implements and methods in the last 100 years than in previous recorded history, and more growth in mechanization in the last 25 years than in the preceding century.

Some changes have been quickly accepted, but others died in infancy. Nevertheless, the continued change and growth in equipment and practices have helped the American farmer become the most efficient food producer in history.

The primary objective of any cropping program is continued profitable production, so most farmers prefer to follow proven practices with readily available equipment. This offers reasonable assurance of predictable results with least risk.

But no tillage operation can be justified merely on the basis of tradition or habit. Any tillage practice that doesn't return more than its cost by increasing yield and improving soil conditions should be eliminated or changed. Contrary

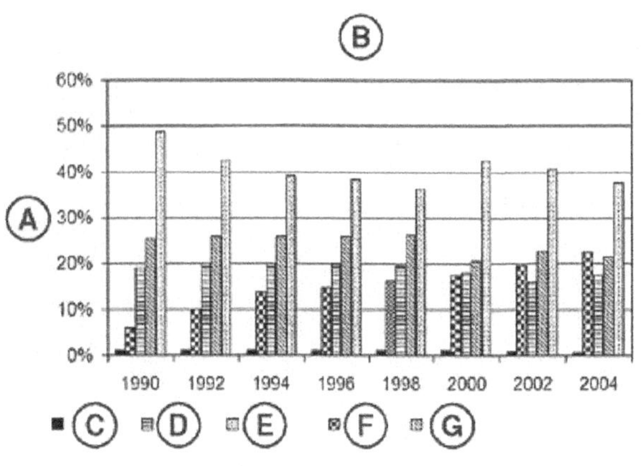

National tillage trends from 1990 to 2004

A—Percent of Planted Acres - All Crops
B—National Tillage Trends 1990 to 2004
C—Ridge-Till
D—Mulch-Till
E—0 to 15% Cover Conventional-Till
F—No-Till
G—15 to 30% Cover

to previous beliefs, soil needs to be worked only enough to ensure optimum crop production and weed control. Any tillage activity beyond that is of questionable value.

Basic concepts and objectives of tillage remain largely unchanged, however the systems employed have moved substantially towards programs of less tillage. Tillage equipment has been developed or reconfigured to meet requirements of these new systems.

Continued on next page

MM16633,0002541 -19-08DEC10-1/2

The greatest change in tillage systems has been a significant shift to conservation farming. This shift has occurred in response to concerns for reducing energy costs, soil erosion, fertilizer and pesticide use, water pollution, and operating costs. The trend to conservation farming has been intensified by federal legislation that can require implementation of conservation practices as a prerequisite for crop support eligibility. Conservation Technology Information Center (CTIC) estimated that conservation tillage was used on 40.7% of all planted acres in the United States in 2004. There is a continued trend upward in the use of conservation tillage. For more information regarding tillage refer to Tillage FMO11104NC.

MM16633,0002541 -19-08DEC10-2/2

Types of Tillage

Tillage can be defined as any mechanical manipulation of soil. Many different types of tillage tools are available to manipulate the soil. A tillage system is the sequence of tillage operations performed in producing a crop. For many tillage systems, the specific operations can be separated into:

- Primary tillage
- Secondary tillage

Primary Tillage

Primary tillage is a deep tillage operation that loosens and fractures the soil to reduce soil strength and to bring or mix residues and fertilizers into the tilled layer. The implements used for primary tillage include moldboard, chisel, and disk plows; heavy tandem, offset, and one-way disks; subsoilers; and heavy-duty, powered rotary tillers. A few different primary tillage tools are shown in the Tillage Equipment chapter. These tools usually operate at least 6 inches (15.4 cm) deep and produce a rougher soil surface than do secondary tillage tools. They differ from each other as to amount of soil manipulation and amount of residue left on or near the soil surface.

Secondary Tillage

Secondary tillage is used to kill weeds, cut and cover crop residues, incorporate herbicides, and prepare a well-pulverized seedbed. Secondary tillage tools include light- and medium-weight disks, field cultivators, row cultivators, rotary hoes, drags, powered and unpowered harrows or rotary tillers, rollers, and numerous variations or combinations of these. They usually operate at depths of less than 5 inches (12.7 cm).

MM16633,00025DF -19-02DEC10-1/1

Tillage Systems

Tillage Practice	Description	Residue
Conventional-till	Full-width tillage that disturbs the entire soil surface. Generally involves plowing or intensive (multiple) tillage trips that may be done before or during planting. Weed control is accomplished with crop protection products and/or row cultivation.	Conventional-till leaves less than 15% residue cover after planting.
Reduced-till	Full-width tillage involving one or more tillage trips to disturb the entire soil surface. Done before and/or during planting. Weed control is accomplished with crop protection products and/or row cultivation.	Reduced-till is less intensive and leaves 15% to 30% of crop residue on the surface.
Mulch-till	Full-width tillage involving one or more tillage trips to disturb the entire soil surface. Done before and/or during planting. Weed control is accomplished with crop protection products and/or cultivation.	At least 30% of the soil surface must be covered with residue after planting.
Ridge-till	The soil is left undisturbed from harvest to planting except for strips up to one-third of the row width. Planting is done on the ridge and usually involves the removal of the top of the ridge. Residue is left on the surface between ridges. Weed control is accomplished with crop protection products and/or cultivation. Ridges are rebuilt during row cultivation.	Residue remains on the surface between the ridges.
No-till, or Strip-till	The soil is left undisturbed from harvest to planting except for strip-till, where strips up to one-third of the row width may receive residue disturbance or may include soil disturbance. Weed control is accomplished primarily with crop protection products. Cultivation may be used for emergency weed control. Other common terms used to describe no-till include direct seeding, slot planting, zero-till, row-till, and slot-till.	Leaves maximum crop residue on the surface.

Table 1 — Tillage Practices

Table 1 — Tillage Practices

Because of the number and variations of the tillage systems used by farmers, it is difficult to give each system a meaningful name or precise definition. The systems can, however, be identified or grouped according to one of the following:

- Overall objective: Names of systems include conservation tillage, conventional tillage, clean tillage, minimum tillage, reduced tillage, and mulch tillage.
- Specific operation: Names of systems include moldboard plow, chisel plow, disk, no-till or no-tillage, and ridge-till.

The name problem is further compounded by the fact that definitions differ between regions of the country. Different names are used to mean the same tillage system, or the same name refers to different tillage systems, all depending on where you live. Therefore, to accurately define or describe a given tillage system, all the operations that make up the system should be listed. The list should include all the tillage operations, any chopping or shredding of residue, application of pesticides and fertilizers, planting, cultivating, and harvesting.

Common tillage practices are shown in Table 1.

Strip-till

Strip tillage, or "strip till," represents a promising soil and residue management system aimed at improving the seedbed environment for early crop growth, preferably in corn, when compared with no-till systems. Several benefits of strip-till are:

- Efficient fertilizer placement
- Seedbed preparation (warmer, dryer soils)
- Increased water infiltration
- Decreased soil erosion
- Reduced compaction
- Moisture conservation in arid climates

Continued on next page

MM16633,00025E0 -19-10NOV10-1/2

Strip-till combines the soil drying and warming benefits of conventional tillage with the soil-protecting advantages of no-till by disturbing only the portion of the soil that is to contain the seed row. This is done by tilling a strip of soil eight to 10 inches wide to prepare a seedbed and leaving the rest of the soil undisturbed. Corn grown in a strip-till system can consistently yield as well as corn grown in a conventional-till system and can yield better than no-till corn crops planted on the same date. The Soil Tillage Intensity Rating (STIR) value for all soil-disturbing activities shall be no greater than 30 to be considered strip-till. Consult your local NRCS County Office for strip-till widths in various row-crop spacings.

Not only is strip-till a conservational type of tillage, but it allows farmers to apply recommended rates of fertilizer in a zone from 6–8 inches deep in the same pass. A simple mole-type knife forms a small tunnel underneath the soil surface, trapping the fertilizer (dry or gas) in a zone for the following crop to utilize. This tillage strip and placement of fertilizer shall be followed by the planter, in the same direction, for best crop utilization.

MM16633,00025E0 -19-10NOV10-2/2

Primary Tillage Equipment

Primary tillage can be further classified as inversion or non-inversion tillage. Inversion primary tillage completely inverts the soil to bury residue. It leaves the soil rough but with little or no surface residue. In contrast, non-inversion primary tillage works the soil to the same depth, but is intended to disturb the surface as little as possible. It leaves the surface rough and covered with most or all of the pre-tillage residue.

Disk

Moldboard plow

Disk ripper

Minimum till ripper

Inversion Primary Tillage

Moldboard Plow

The moldboard plow cuts, lifts, and turns the furrow slice and in so doing it:

- Buries some or all of trash and crop residue
- Aerates the soil
- Controls weeds, insects, and crop diseases
- Incorporates fertilizer into the soil
- Prepares the soil for establishing a good seedbed

Incorporation of plant residue and aeration of the soil by tillage also stimulate growth of microorganisms, which decompose trash and other organic materials. Accelerating decomposition of organic matter increases the supply of available nitrogen, phosphorus, potassium, and other plant nutrients. Soil microorganisms require the same favorable conditions of warmth, moisture, and aeration as are needed for quick seed germination.

Good plowing sandwiches organic matter between furrow slices to form a wick for better water absorption and storage and for faster decomposition of residue. It also increases soil porosity and provides more air for faster, stronger root growth.

Good soil pulverization in plowing reduces the cost of later tillage, but unless planting will start immediately, each furrow should retain a distinct crown. This helps

Plowing buries trash and aerates the soil

reduce runoff, reduces ponding and puddling in low spots, reduces wind erosion, increases moisture infiltration, and speeds surface drying.

Plowing to the same depth each year may lead to a plow sole or hardpan, just below plowing depth, which can severely restrict root growth and water movement. Occasional chisel plowing or plowing several inches deeper every few years will help break the compacted layer.

Continued on next page

MM16633,00025A3 -19-10NOV10-2/5

Non-Inversion Primary Tillage

Non-inversion primary tillage, unlike inversion tillage, is intended to disturb the surface as little as possible and to leave it rough and covered with most or all of the residue present before tillage. This section will discuss the most popular tools now used for non-inversion primary tillage.

Chisel Plows

The basic function of a chisel plow has changed little from that of the forked stick pulled through the soil by primitive man thousands of years ago. Alloy steels have replaced the wood and tractors have replaced animal and human muscle power, but the purpose still is to stir and aerate the soil with little inversion. Using today's wider chisel plows that can till at high speeds, one person can till hundreds of acres per day.

Chisel Plows vs. Moldboard Plows

Draft of a chisel plow is perhaps half that of a moldboard plow per foot of width — both working the same depth. Therefore, chiseling is faster and more economical than moldboard plowing where complete trash coverage is not required. However, if soil is chisel plowed twice to prepare a seedbed, more fuel may be used than would be needed for moldboard plowing. Chisel plows are also frequently used to break up the hardpan formed from years of plowing at the same depth with a moldboard plow.

Because chisel plows break and shatter the soil, they perform best when soil is dry and firm. When too wet, soil is merely split open by the shank with no shattering or pulverizing. In fact, if soil is chiseled when it is too wet, large clods may be formed that are difficult or nearly impossible to break up with subsequent tillage to form a suitable seedbed.

Wide chisel plows can work hundreds of acres a day

Chisel Plows vs. Field Cultivators

Although chisel plows are referred to in some areas as field cultivators, we are classifying them here as distinct machines. Chisel plows have heavier construction and are basically used for primary tillage. Field cultivators are used principally for secondary tillage, weed control, and seedbed preparation. Field cultivators are much lighter in construction than chisel plows, and are designed for shallower operation.

Modern chisel plows normally have two or more rows of curved spring-steel shanks attached to a rugged box-steel frame. The shanks are arranged in staggered rows to permit better trash flow and laterally balance the draft load of the machine.

Continued on next page

MM16633,00025A3 -19-10NOV10-3/5

102615
PN=31

Heavy-Duty Disks

Heavy-duty disks are used for primary tillage such as tilling unworked soil, cutting and mixing heavy crop residue and mulching stubble. They are often used in fields with cornstalks or other residue, ahead of plowing. This loosens the surface, cuts residue, and mixes it in the soil. In clean tillage systems, this operation provides better residue coverage when the land is later plowed, more soil for better soil-residue contact, and faster decomposition of residue.

Heavy-duty disks are used for primary tillage

MM16633,00025A3 -19-10NOV10-4/5

Subsoilers or Rippers

Subsoiling or ripping usually is done to loosen or break up impervious soil layers below the normal tillage depth to improve water infiltration, drainage, and root penetration. To be effective in improving crop yields, subsoiling must meet these conditions:

1. It should be done when soil is relatively dry for maximum effect on the hard layer. If soil is wet, only a thin slot is sliced through the soil, which will likely reseal very quickly, and down-pressure of tractor weight and subsoilers can cause compaction.
2. Soil below the impervious layer must have excess water-holding capacity, or there will be no place for surface water to go, and no air in the deeper layers for plant-root growth.
3. Deeper soil must not be so acid or alkaline as to discourage root growth.
4. Tractors and heavy implements must run at least 1 foot (30 cm) away from subsoil slots during subsequent operations, to prevent resealing of the slot by tire compaction.

Some outstanding results have been achieved from subsoiling. Yield increases of 50% to 400% have been

Subsoilers break hardpan

reported from subsoiling under the right soil and moisture conditions and in the right areas. But, in other cases, little effect has been observed.

Depending on subsoiler design, power available, soil conditions, and depth of hardpan, subsoiler penetration may be as deep as 24 inches (60 cm).

MM16633,00025A3 -19-10NOV10-5/5

Secondary Tillage Equipment

Secondary tillage operations, less aggressive than primary tillage, are used to prepare fields for planting, for summer fallowing, chemical incorporation, and weed control. The degree of tillage varies with the type of tillage system. With a conservation tillage system, one pass primarily to reduce surface roughness or kill small weeds might be sufficient. A clean-tillage system may require two or more passes with more than one tillage tool to provide a well-tilled surface free from residue and weeds. Light and medium disks, field cultivators, combination tillage implements, and various harrows and packers are commonly used in secondary tillage operations.

Rotary hoe

Row-crop cultivator

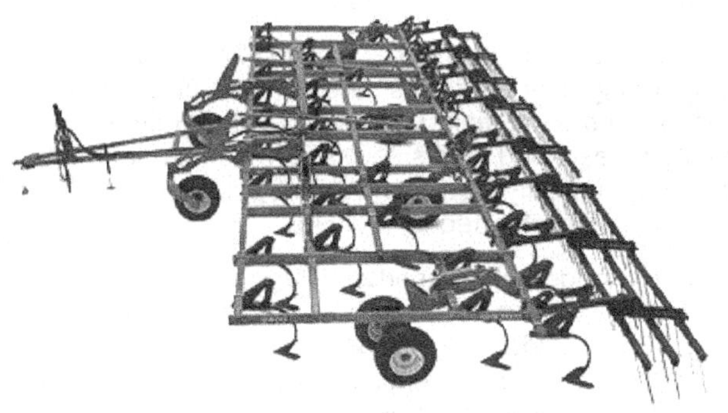

Field cultivator

Combination tool

Continued on next page

MM16633,00025E2 -19-08DEC10-1/3

Disks

Light to medium disks, or properly adjusted and equipped heavy-duty models, are well suited for secondary tillage. With the rapid adoption of conservation tillage systems, disks are currently used more frequently for secondary than for primary tillage.

Disks used primarily for secondary tillage are available as integral or drawn types. Integral equipment is attached to the tractor 3-point hitch or implement quick-coupler, and the entire implement is carried by the tractor during transport. A drawn disk is a complete unit in itself attached to the tractor drawbar, and is raised or lowered by hydraulic cylinders. Operating widths vary from about 6 feet (1.8 m) to 32 feet (9.8 m) in folding models. Furrow filler, tine-tooth harrow, and finishing blade accessories add to the capacity of disks to provide a finished smooth seedbed.

Drawn disk

MM16633,00025E2 -19-08DEC10-2/3

Field Cultivators

Field cultivators are widely used across North America for seedbed preparation, weed control, stubble-mulch tillage, summer fallow, and roughing fields to increase moisture absorption and control wind and water erosion.

Field cultivators and chisel plows have similar appearance and operating characteristics, and in some areas chisel plows are known as field cultivators. But the field cultivators described in this section are constructed and used for lighter-duty field operations than the deeper primary tillage performed by a chisel.

They are intended for secondary tillage of previously worked soil, such as seedbed preparation in fall- or spring-plowed fields, use after stalks or stubble have been chisel plowed or disked, summer fallowing after use of a disk tiller, chisel plow, or wide-sweep plow, and similar operations.

Field cultivators leave most of the residue on top or mixed into the upper few inches of soil depending on the type of sweep or point that is attached to the shank. When using spike points, the surface is usually left ridged, rough, and

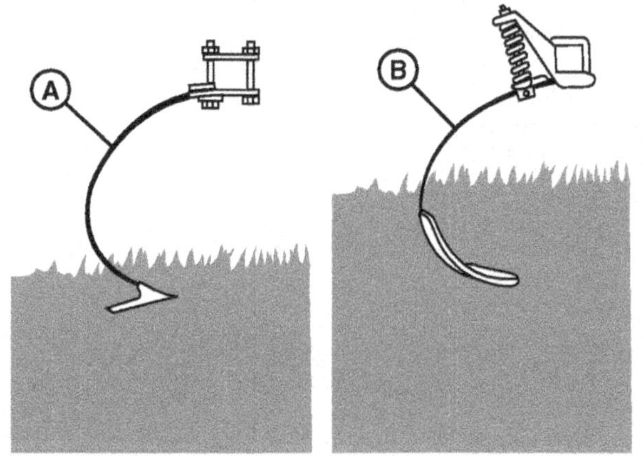

Field cultivators are intended for less-severe conditions

A—Field Cultivator   B—Chisel Plow

open so moisture infiltration is increased and soil blowing and runoff are reduced.

MM16633,00025E2 -19-08DEC10-3/3

Tillage at Seeding

Conservation tillage systems are designed to retain a considerable amount of surface residue from the previous crop. Planters, drills, and seeders must open furrows and place and cover seeds while leaving the soil surface as undisturbed as possible. Some tillage, therefore, is required at seeding. With no-till systems, no prior tillage is done except for a narrow strip just ahead of the opener that is made by a coulter or tool attached to the planter or seeder.

Tillage at planting enables the farmer to reduce the number of trips over the field. Also, shortage of farm labor and larger farm sizes have increased the popularity of combining tillage and planting in both full-tillage and conservation farming systems.

Heavy-duty row-crop planters, grain drills, and seeders have been developed specifically for satisfactory planting in high residue or no-till conditions. Tillage attachments are also available for conventional planting equipment that provide for planting in many tilled and untilled conditions.

Row Crop Planters

Modern row-crop planters are designed for performance in seedbeds that range from clean-tilled to heavy-residue, conservation-type conditions. Hinged and folding frame designs permit consistent planting over terraces and contours and also narrow transport width of wide planters.

Several types of tillage attachments are available to improve planter performance in tilled or untilled seedbeds:

- Disk furrower attachment
- Tru-Vee opener
- Tine-tooth tillage attachment
- Fluted, ripple, and serrated coulters
- Row cleaner attachment

Row-crop planters can operate well in both clean-tilled and conservation seedbeds

Continued on next page

MM16633,00025A4 -19-21OCT10-1/4

A common practice in some areas is to run a planter behind a field cultivator on fall-plowed land. The disk furrower attachment pushes clods from in front of the furrow opener to provide more uniformity in planting depth.

Disk furrower attachment

MM16633,00025A4 -19-21OCT10-2/4

Tru-Vee openers slice a narrow, V-shaped furrow creating a prime environment for seed. The two disk blades easily carve through hard soil and heavy residue.

Tine-tooth tillage attachments prepare only the seedbed or row area.

Tru-Vee opener attachment

Continued on next page

MM16633,00025A4 -19-21OCT10-3/4

Fluted, ripple, and serrated coulters are no-till planter attachments that help prepare the seedbed without prior tillage. The fluted coulter prepares a seedbed about 3 inches (7.5 cm) wide and 2 to 3 inches (5 to 8 cm) deep.

The ripple coulter prepares a seedbed about 0.75 to 1.5 inches (1.9 to 3.8 cm) in width.

Surface residue in front of the planter can hairpin at the disk opener and cause poor soil penetration. The row cleaner attachment is designed to sweep away the residue for improved opener performance.

Fluted coulter for no-till planter

Row cleaner attachment

MM16633,00025A4 -19-21OCT10-4/4

Post-Emergence Tillage

Weed control, preparation of land for irrigation, and breaking soil surface crust are primary purposes of post-emergence tillage.

Sweeps, shovels, and rotary-cultivator or rotary hoe wheels slice off weeds, pull small weeds, and shatter the soil surface to expose additional weed roots to die in the sun. When using small sweeps and shovels, some well-rooted weeds are apt to slip around the tool without being cut. This can be minimized by overlapping large sweeps or weeding knives, and must be considered in arranging tools on the cultivator.

Weeds usually germinate in the upper 1 to 2 inches (25 to 50 mm) of soil, so shallow cultivation will normally control those weeds and reduce germination of additional seeds that are buried deeper. High concentration of soybean feeder roots have been found in the top 2 to 4 inches (50 to 100 mm) of soil. Corn roots will quickly spread across the entire inter-row area in uncompacted soil. Cutting the crop's roots with deep cultivation reduces the its ability to gather moisture and nutrients from the soil and can seriously reduce yields.

Soils that tend to crust seem to benefit most from cultivation. Stirring the soil surface breaks up the crust, improves aeration and moisture absorption, and kills weeds growing between rows. Cultivation also helps control weeds that may be resistant to herbicides, and may discourage germination of some late-starting weeds, such as fall panicum.

Equipment typically used for post-emergence tillage includes front-, mid-, and rear-mounted row-crop cultivators, rotary cultivators, and rotary hoes.

Rear-mounted row-crop cultivator

Rotary cultivators can work up to 8 miles per hour (13 km/h)

Rotary hoe provides fast early cultivation

MM16633,00025A5 -19-30SEP10-1/1

Toolbars

Toolbar tillage and planting began in the Southwest and later moved into the Mississippi Delta area. Only in recent years has there been much interest in toolbar equipment in the Corn Belt. With relatively small tractors and stable 40-inch row spacing (102 cm), there formerly was little incentive for Corn Belt farmers to use toolbars.

But as row spacings began to change, and planting and cultivation became 6-, 8-, 12-row, and even wider operations, toolbars have taken on new importance in assembling the particular equipment each farmer needs. Since the late 1960s the variety, size, and use areas of toolbar equipment have grown rapidly.

IMPORTANT: Every tractor has certain lift and draft limitations, and all ready-made equipment is likewise designed and constructed to meet definite strength and performance specifications. This permits manufacturers to provide recommendations for matching of implements to specific tractor models.

However, most toolbar-based implements assembled by dealers or farmers have not undergone the extensive testing of factory-built machines. Therefore, it may be difficult to predict the exact size of bar required, the correct number of clamps, or the lift capacity required to transport the assembled machine. Too much weight can overload the tractor hydraulic system, reduce tire life, and cause poor tractor stability. Excessive draft means unsatisfactory field performance and possible tractor or equipment damage, and could cause tractor instability.

Too much power for the strength of the implement could result in serious equipment damage, especially in adverse operating conditions.

The best method for determining toolbar size and tractor capacity is to compare plans for combinations with similar toolbar implements offered by the manufacturer for certain tractor models. It also is helpful to check with farmers who may be successfully using an assembly similar to what is being planned.

For many years toolbars were solid-steel bars, 2, 2-1/4, or 2-1/2 inches (50, 57, or 64 mm) square. Later, square hollow bars became available to reduce weight by 55% to 60%, cut cost by as much as 20%, and retain ample strength for many small jobs. These also are 2 to

Toolbar equipment matches individual needs

2-1/2 inches (50 to 64 mm) square and often are used interchangeably or in combination with solid bars.

A hollow 3-1/2 x 7 inch (89 x 178 mm) flattened-diamond shape is used for some toolbars. This shape provides considerably more strength than small, square toolbars, and permits use of some of the same attachments by using longer bolts.

Many rear-mounted row-crop cultivators use 4 x 4 or 5 x 7 inch (100 x 100 or 127 x 178 mm) hollow bars for maximum strength per pound. However, these bars are seldom used for other applications, so they will not be discussed further here.

Many rear-mounted row-crop cultivators use 4 x 4 or 5 x 7 inch (100 x 100 or 127 x 178 mm) hollow bars for maximum strength per pound. However, these bars are seldom used for other applications, so they will not be discussed further here.

The type of toolbar must be decided on the basis of anticipated use and total weight. Oversized toolbars provide added strength, but increase weight and cost. However, replacement of toolbars that fail due to insufficient strength may increase total costs even more.

MM16633,000258A -19-30SEP10-1/1

Soil-Engaging Components

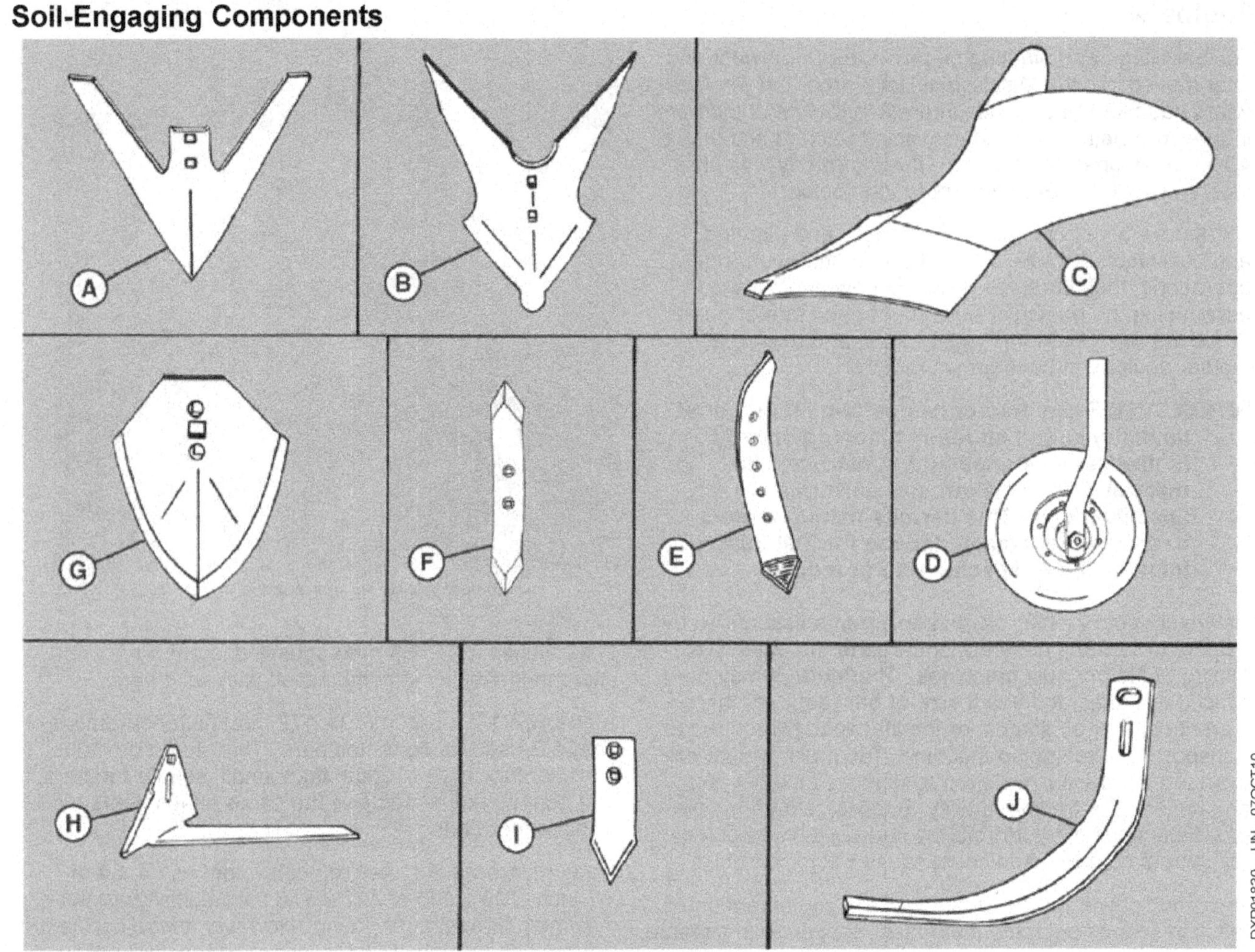

Soil-engaging tools

A—High–Crown Sweep
B—Furrow Opener
C—Lister Bottom

D—Weeding Disk
E—Twisted Shovel
F—Reversible Chisel Point

G—Heavy–Duty Furrow Opener
H—Square–Turn Knife
I— Spear–Point Shovel

J— Curved–Bed Knife

Tillage equipment may be equipped with a variety of soil-engaging tools to match almost any soil, crop, or tillage condition. Options include chisel-plow sweeps, shovels, furrow openers, reversible spike and chisel points, and twisted shovels. Also available, with different standards, are various sweeps, points, shovels, disk hillers, furrow openers, and weed knives.

Continued on next page

MM16633,000258F -19-09DEC10-1/2

Soil Texture

Soil-engaging components (e.g., sweeps, shovels, chisel points, etc.) wear out over time because they run in the ground. Several factors affect the speed of wear, some of which include:

- Soil texture
- Speed of operation
- Depth of operation
- Soil compaction
- Soil moisture content
- Previous tillage practice
- Soil crusting

Soil is composed of three basic particle sizes: sand, silt and clay. Particles of sand range from 0.05 to 2.0 mm diameter, while silt particles range from 0.002 to 0.05 mm. Clay particles are the smallest at less than 0.002 mm. Soils high in sand will be more abrasive to metal components and cause higher wear than silt or clay soils.

Soil is classified by its relative proportions of sand, silt, and clay particles. The specific percentage of each of the three particle sizes determines the soil's textural class.

- Soil textural classification provides a more general approach to determining the characteristics of a soil in a given area. Each textural class has its own abrasive characteristics which impacts wear on tillage equipment.
- Soils with high amounts of sand in their composition tend to be gritty to the touch. Sandy soils have a coarse texture and are composed of abrasive quartz and silicate minerals.
- Clay soils tend to have a fine textural composition and are somewhat "sticky" to the touch. Tillage operations

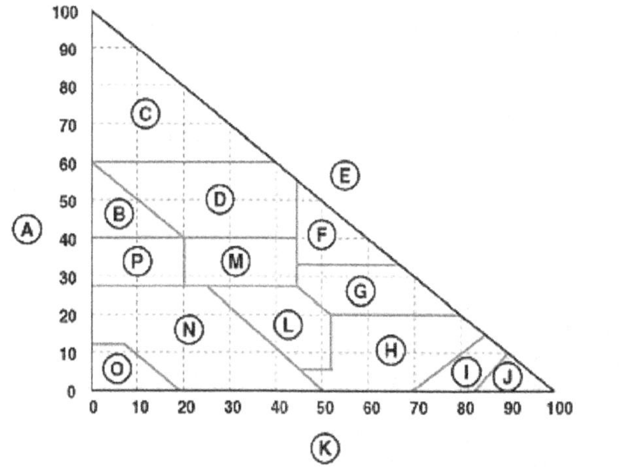

Soil texture classification

A— % Clay	I— Loamy Sand
B— Silt Clay	J— Sand
C— Heavy Clay	K— % Sand
D— Clay	L— Loam
E— Textural Classes	M— Clay Loam
F— Sandy Clay	N— Silt Loam
G— Sandy Clay Loam	O— Silt
H— Sandy Loam	P— Silty Clay Loam

in high-clay soils will provide normal rates of wear, but are dependent on other factors, such as moisture.
- Silt soils have moderately-coarse to moderately-fine textures. Because silt particles are larger than clay particles, they can be more abrasive on tillage tools.

MM16633,000258F -19-09DEC10-2/2

Tractor and Implement Preparation and Adjustment

Both the tractor and the tillage tool that it pulls must be properly set up for maximum tillage efficiency and desired performance. Adjustments to attain these objectives are related primarily to traction, flotation, and soil compaction. These factors are closely related, and changes in soil-surface conditions, soil contact area, and vehicle weight will directly affect all three.

Traction is the linear force, pull or draft, resulting from torque applied to tractor tires. Traction is usually thought to be good when wheel slippage is not excessive.

Flotation is the ability of tires to stay on top of the soil surface or resist sinking into the soil.

Both traction and flotation are directly related to vehicle weight, soil conditions, and contact area between the tire and soil surface.

Compaction is the packing or firming of soil caused by wheel traffic. It is usually undesirable because it can restrict movement of air, water, and crop roots in the soil.

Several means are available to increase traction and flotation while reducing the potential for soil compaction.

Tractor Setup

Tractors are built with ample size and power for specific farming operations. But, to get maximum benefit from that power, additional weight may be required to gain maximum drawbar pull and sufficient traction for high draft tillage implements.

Adding ballast (weight) to drive wheels and tractor front end is the most common method of improving traction and helping increase drawbar pull.

Front-end weight may also be required for tractor stability, particularly with integral and heavy-draft, semi-integral equipment.

A tractor will provide best performance when it is equipped with tires that are large enough relative to tractor weight to give a soft ride (low stiffness). In most situations, tires should be inflated to operate at rated deflections, i.e., minimum pressures to support static weight carried on each axle. The larger the tire air volume the better. For

Weight transfer improves traction—part of front-end weight is transferred to rear wheels; part balances rear loads

radial tires, a rule of thumb is that the tires should require no more than 14 psi (0.97 bar), preferably less, to support the ballasted static axle load. Best results will be obtained when there is more tire on the tractor than might have been thought necessary. A tractor being set up for heavy tillage work should be considered in terms of bigger tires and moderate to light weight rather than smaller tires and heavy weight.

Three major items to be considered are:

- Total tractor weight and static weight split (weight distribution on front and rear axles)
- Type of ballast used (cast and/or liquid weight)
- Tire inflation pressures

None of these can be considered independent of the others to achieve optimum tractor performance. In addition to improved traction (reduced slippage and higher fuel efficiency) and improved flotation, a properly set-up tractor will provide the following benefits:

- Reduced compaction
- Improved ride
- Reduced tire wear
- Improved sidehill stability
- Better control of power hop on mechanical front wheel drive (MFWD) and four-wheel drive (4WD) tractors.

MM16633,000258B -19-08DEC10-1/1

Introduction

Planting or seeding of agricultural crops was one of the earliest farming operations to be mechanized as the agricultural revolution moved across America. The modern farmer uses precision planting/seeding equipment and the highest quality "certified" seeds. Planting at the correct time, preparing and maintaining soil conditions, and the efficient use of farming equipment helps today's farmers achieve maximum yields.

MM16633,00025A6 -19-02DEC10-1/1

Selecting Good Seeds

The modern crop farmer realizes the importance of using quality seed. The seed variety, the germination percentage and the purity of seed (free of inert matter) should be known before it is planted. The crop farmer cannot harvest a better crop than the seed planted; therefore, the use of good seed is the first step in increasing crop yield and income.

The characteristics of good seed includes:

1. Seeds that will produce plants adapted to the climate and soils where grown.
2. Seed that is consistently good quality year after year.
3. Seed that is pure.
4. Seed that is free of weed seed and disease.
5. Seed that will germinate.

Quality seed is the first step to high yield

Continued on next page

MM16633,00025E4 -19-23NOV10-1/4

The seed tag indicates the quality of the seed

A—Variety Number	D—Front of Seed Tag	G—Warm Germination
B—Date of Testing	E—Purity	H—Type or Shape of Kernel
C—Back of Seed Tag	F—Origin of the Seed	

One of the ways to help ensure a crop is to start with good seed. Much time and money goes into the development of seed that will produce high yields.

Because the appearance of seed tells little about its quality, seed producers are required by law to mark each bag of seed with a seed tag. This is your assurance of quality and of what you might expect from the seed.

The seed tag tells the following:

1. Warm germination
2. Purity
3. Variety number
4. Origin of the seed
5. Date of testing
6. Type or shape of kernel

Continued on next page

MM16633,00025E4 -19-23NOV10-2/4

The back of the tag shown has a notice regarding the conditions of liability or responsibility that the seed company has to the buyer. The tag also gives the buyer precautions regarding seed treatment with poisonous or dangerous chemicals. These notices must appear on each tag to protect both the buyer and the producer.

MM16633,00025E4 -19-23NOV10-3/4

Seed Germination

While selecting quality seed is an important first step, unless the proper external conditions are provided for germination and growth, plants will not grow to maturity and produce a crop.

Three basic environmental conditions are necessary for germination of matured seed:

- Moisture
- Oxygen
- Heat

These conditions must be present in the proper proportions for top seed germination.

Genetically Modified Crops

Starting with the development of hybrid corn seed in the 1920s, improvements in varieties have grown rapidly as seed companies improved not only the seeds, but also their plant breeding techniques. Once gene-transfer techniques were perfected, allowing genetic materials from one organism to be copied to another, unlimited new possibilities emerged for crop production. In the 1990's the first genetically modified (GMO) crops offered

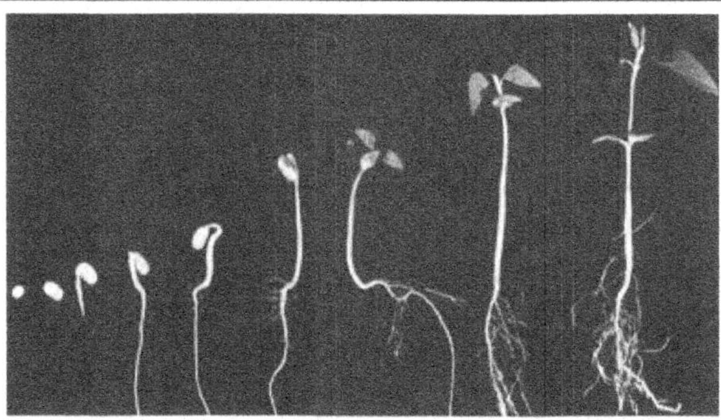

Good germination of seed is necessary for producing high yield

either resistance to specific, major insect pests (Bt corn and cotton), or tolerance to glyphosate herbicide (Roundup-Ready™ soybean). Many other traits that either protect the crop, improve its suitability to a specific end-use (such as altered oil content), or simply improve its durability (drought-tolerance) are constantly being studied and presented to the market. Approval of these specific traits by regulatory agencies as well as export markets is necessary before the new variety can be brought into the commodity marketing systems.

MM16633,00025E4 -19-23NOV10-4/4

Functions of Planters and Grain Drills

The purpose of most planters and grain drills (excluding broadcast planters) is to plant seeds evenly in rows or on beds. To do this in the manner desired, the planter must perform a number of important functions:

- Open a furrow in the soil
- Meter the seed
- Place the seed
- Cover the seedbed
- Firm the seedbed

The devices which perform these functions are briefly discussed here; their operation will be described in more detail in later sections of this text.

Types of Planters

Planting or seeding equipment is generally divided into four types:

- Row-crop planters
- Grain drills and air seeders
- Broadcast seeders
- Specialized planters

A number of different types of equipment for planting of different crops are available in each of the categories.

MM16633,00025E3 -19-30SEP10-1/1

102615
PN=45

Row-Crop Planters

Sizes of Row-Crop Planters

Row-crop planters are sized according to the number of rows they will plant and the space between rows. While planters with 12, 16, and 24 rows are the most common sizes, planters with 4, 6, 8, 36, and 48 rows are available. The most common row width for corn is 30 inches (76 cm), but particularly in more southern regions, some 36 and 38 inch (91 and 97 cm) row widths are planted. Northern states have been adopting 20 and 22 inch (51 and 56 cm) planters for corn.

Another planter configuration that is becoming very popular is the split-row planter, which will plant 30 inch (76 cm) row widths for crops such as corn, but can also plant 15 inch (38 cm) rows for soybeans, which favor narrow row spacing.

Options and Features for Row-Crop Planters

To meet the varied needs of farmers and their crops, manufacturers offer a wide variety of features on planters.

Frame Configuration — Planter frames, as with many implements, may be designed as either drawn or integral. Because of its firm connection with the tractor, an integral planter is often preferred when accurate tracking is essential, such as planting cotton on beds. Drawn frames are available in much larger configurations than an integral tool can accommodate. Due to their large size, the planter frame may be folded in order to transport or store the machine, for example by stacking vertically, folding up wings, or folding vertically. "Split-row" planter frames fold or row units can be brought up to quickly transform a planter from 15-inch row spacings to 30-inch or 20-inch to 40-inch.

Central Fill — Timeliness is crucial during the planting season, and as planter sizes have grown to as many as 48 rows, central-fill planters have emerged as a time saving alternative to filling the individual seed boxes on each row unit. Large seed hoppers, similar to a seed tank on an air seeder, hold and deliver the seed to the meters on the row units.

Meters — Meters in row-crop planters "singulate" or separate the seed by means of either air, vacuum, or mechanically as with a finger-pickup meter. In contrast, drills and air seeders typically use volumetric meters that deliver a measured volume of seed instead of precisely separating and placing individual seeds. Once the meter has captured the seed, it drops it down the seed tube, where it is placed in the soil with the opener.

Openers — Unlike air seeders which can utilize a variety of openers to create a seed furrow, most row-crop planters are equipped with double-disk openers. Double-disk openers can precisely place seeds with guidance to correct and consistent depth by the gauge-wheel.

Chemical Application — Advancements in seed technology that provide insect resistance have reduced

Twelve-row front-folding planter

the need for soil insecticides to be applied with the planter. Nonetheless, new insect pests or special situations will always create a demand for insecticide delivery systems. Pesticide manufacturers have developed advanced, closed-system units that reduce operator exposure and minimize the amount of product that needs to be carried or refilled on each row unit. The pesticide is delivered from the storage container through a tube which drops it into or on top of the seed furrow. Liquid pesticide systems are available which reduce pesticide exposure and specialized application equipment, and increase convenience.

Similarly, both dry and liquid fertilizer applications may be made in or along the seed furrow. Small amounts of fertilizer may be dropped in the seed furrow along with the seed to feed early growth of the seedling. If the fertilizer is applied with a separate opener up to 2 inches below and on the side of the seed row, larger amounts of fertilizer can be safely applied without damage to the seed.

Coulters and Row Cleaners — Due to increased adoption of no-till and conservation-tillage systems, many planters are equipped with attachments to handle heavy residue. Row cleaners sweep residue away from the seed row to allow less interference with the opener, plus allow for faster warming of the soil. Coulters can slice through residue that intersects the seed row, and till a thin band of soil just ahead of the seed opener.

Precision Farming — New technologies joining precision farming and planting present many opportunities to improve efficiency and productivity. Automated guidance systems not only eliminate the need for row markers, but also offer more efficient coverage of the field. Improved seed monitors continually check for accurate seed metering and delivery, immediately alerting the operator if malfunctions occur. Malfunctions in seed metering and delivery may otherwise be very difficult to discover until the crop emerges. Planters equipped with row- or section-control can engage or disengage individual row clutches to eliminate over planting on point rows, head rows, and on irregularly shaped fields.

MM16633,00025A7 -19-09DEC10-1/1

Grain Drills

There are three major types of grain drills:

- End-wheel drill
- Press-wheel drill
- No-till drill

End-wheel drills have wheels that support and drive the drill. A press-wheel drill has press-wheel gangs mounted on the rear of the drill. The press wheel firms the soil over the seed, drives the metering mechanisms, and supports the rear of the drill. A yoke and wheel are used to support the front of the drill.

Drills may be purchased as a plain drill which has seed hoppers only or as a combination drill which has a seed hopper and fertilizer hopper. Fertilizer attachments may be added to plain drills if offered separately by the manufacturer.

Because drills are used to seed crops into narrow rows of 6 to 15 inch spacings, a well prepared seedbed is the ideal situation to keep residue from interfering with the closely-spaced drill openers. However, many of the crops that are seeded with drills, such as small grains, soybeans, hay or pastures, also lend themselves well to no-till opportunities. No-till drills were introduced in the late 1980s which offered heavier frames, better residue clearance, coulters for cutting residue, and down-pressure to penetrate both residue and untilled soils. These seeding tools were instrumental in the very rapid adoption of no-till through the 1990s.

Grain Drill Size

Grain drills are usually sized according to the number of furrow openers and the distance between openers. They may also be sized by drilling width in feet. An example of the more common method of drill sizing is 17 x 7. In this case, 17 furrow openers are spaced 7 inches apart. The width of this drill would be 17 x 7 = 119 in., or 119 ÷ 12 = 9 ft., 11 inches (3 m).

End-wheel drills may have 12 to 24 openers. Typical spacings are 6 to 10 in. (15 to 25 cm).

End-wheel drill

No-till drill

Press wheel drills may have 6 to 24 openers. Spacings may be from 6 to 16 in. (15 to 40 cm) apart.

Some drills with narrow row spacings will have the openers staggered: every other opener being in one rank and the alternate openers in a second or third rank. This arrangement permits greater residue clearance than if the openers were in a straight line.

MM16633,00025BB -19-10NOV10-1/1

Seeders

Air Seeders

Air seeders were developed in Germany over 20 years ago for the purpose of seeding small grain. The concept has become popular in areas where grain is produced on large flat areas of land, such as the wheat areas of Australia and Great Plains of the United States and Canada.

The seeders consist of two separate implements (the seed tank and a seeding tool) operating together to perform as one seeding unit. This combination allows the field operations of tilling, seeding, and fertilizing to be accomplished in one field pass. The opener on the seeding tool can be a single disk, double disk, or a hoe-type opener.

Air seeders feature an air-delivery system to assure uniform seed distribution from the seed-metering mechanism to the seedbed. The air stream is the most efficient method of transferring the seed and fertilizer from central seed tanks to a number of remote locations on the chisel shanks.

Central seed tanks provide greater capacity and, therefore, results in fewer stops to refill the tank.

Additional advantages of this system include:

- Central fill tank and central metering of grain and fertilizer
- Few moving parts providing low maintenance
- Reduced tillage requirement prior to seeding
- Higher operating speeds
- Good residue and rock clearance
- Easy and fast road transport

Broadcast Seeders

A second method for the solid planting of crops is to use a broadcast-type seeder.

An air seeder combines tilling, seeding, and fertilizing in one field pass

Broadcast seeders may be either the centrifugal-type spreader or the full-width-feed broadcaster, also called the field distributor.

Broadcast seeders do not have furrow openers: therefore, the seedbed must be totally prepared by a tillage tool such as a disk harrow. Also, broadcast seeders usually do not cover the seed. Seed covering may be done later with a spike-tooth harrow or similar piece of equipment. The use of a harrow or very minimal tillage application must follow this type of seeding to cover seed and allow seed germination.

MM16633,00025E5 -19-11NOV10-1/1

Planting to Maximize Yields

Inputs for Optimum Yields

Several factors to consider are:

- Time of planting
- Seeding depth
- Soil preparation
- Soil fertility
- Tillage
- Weed and insect control

Let's look at how these factors affect the crop yield.

Time of Planting

The best time of planting varies with the crop and the geographic location. Crops yields are decreased if planting is done too early or too late. Early planting of crops such as corn and soybeans has these advantages:

1. Getting an early start can mean more time is available for seedbed preparation and planting. The workload can be spread over a longer period of time if necessary.
2. Harvest may also be started earlier and spread over a longer period of time if early varieties are used.

If planting is done too early, cold weather may damage the seedling and this would require replanting. Corn will germinate quickly when the soil temperature reaches 55° to 60°F (13° to 16°C). Soil temperatures reach this range when the highest daytime temperatures are 65° to 70°F (18° to 21°C).

Thermometer used to measure soil temperature

The soil temperature should be checked between 7 and 8 AM. Push a thermometer 3 inches into the soil and leave it for about 3 minutes. If the thermometer reads 50° to 52°F (10° to 11°C), the soil is warm enough to start planting.

Continued on next page

MM16633,00025A8 -19-11NOV10-1/5

Corn and soybeans will stand cold weather as long as the growing point in the plant does not freeze. A corn plant may be frozen off and still live if the growing point has not yet reached the soil surface. The corn plant is usually 12 inches (31 cm) tall and has the fifth or sixth leaf when the growing point reaches the surface of the soil. Frost damage after this stage will kill the plant. Corn is more hardy than soybeans.

The growing point of soybeans is located above the surface of the soil at a point between the two leaf stems. Because the growing point emerges much earlier in soybean than corn, it can be more vulnerable to frost. As long as buds at the point where the cotyledons attach to the stem have not been damaged, the soybean plant can still recover from early frost damage.

A—Growing Point (Freezing   B—1"
    Here Kills Plant)

Location of corn plant growing point

Location of soybean plant growing point

Continued on next page

MM16633,00025A8 -19-11NOV10-2/5

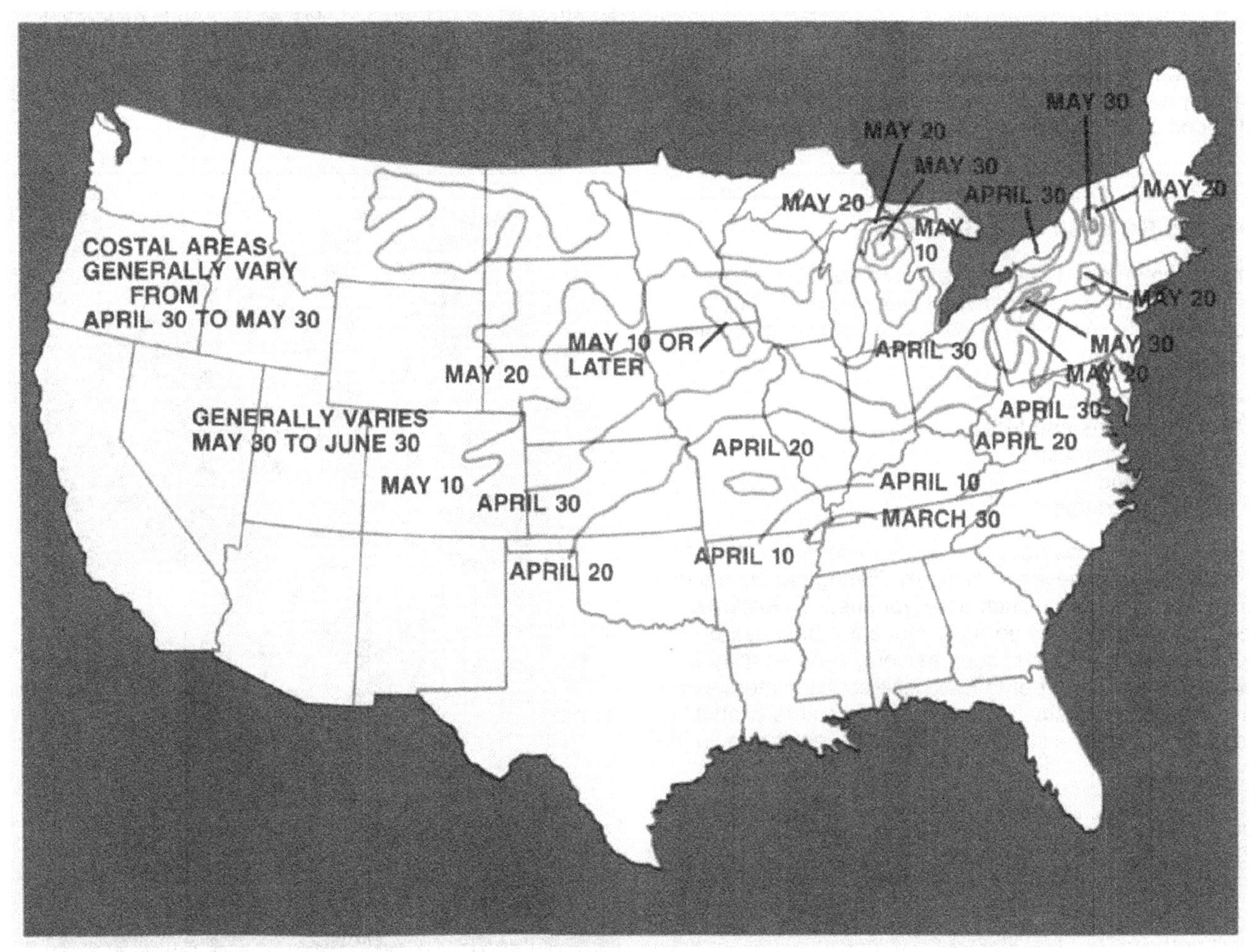

Average date of last freeze

There is a 50% chance that the last freeze will occur after the average date of last freeze given in the above graphic.

To determine the earliest planting date for corn, first find the average date of the last freeze for your area. Add eight days to this to reduce the chances of a killing frost to 25%. (If desired, add 15 days to reduce chances to 10%.)

Next, subtract the number of days required for corn seed to germinate and the plant to emerge, usually between 11 and 15 days. Finally, subtract the number of days required for the growing point to reach the surface, approximately 10 to 12 days.

Here is an example of determining earliest date of planting using this procedure for the central Illinois and southern Iowa region:

1. Average date of last freeze — April 30.
2. To reduce chances of a freeze to 10%, add 15 days to April 30. The new date is May 15.
3. Now subtract days for germination (13 days average) and days required for growing point to reach the surface (11 days average).

4. Earliest date of planting for this region: May 15 minus 24 (11 + 13) days equal April 21 as the day to plant with only a 10% chance of freezing.

Early Planting Compared to Late Planting

Research shows that corn planted after the optimum planting date will have, as a general rule, a reduction in yield of about one bushel per acre for each day of delay in planting. Corn planted after this time does not use effectively the long sunny days of June and early July. And there is a greater chance of heat stress during pollination, which can reduce yield significantly.

The trend is to plant soybeans as early as possible, making sure there is no chance for a killing freeze. Usually soybeans are planted as soon as possible after the corn is planted. For example in the central Illinois-southern Iowa region, soybean planting begins around May 10 to 15.

Remember, the soil temperature must be at least 50° F before the seeds will germinate.

Continued on next page

MM16633,00025A8 -19-11NOV10-3/5

Seeding Depth

Along with the importance of planting seed at the depth which provides adequate moisture for germination, it's also critical that seeds be placed at a consistent depth, so emergence of plants occurs as uniformly as possible. In addition, it's important to follow recommended depth ranges for the crop so that the plants are adequately anchored, but shallow enough for unhindered emergence.

MM16633,00025A8 -19-11NOV10-4/5

Seeding Rate and Seed Spacing

The recommended seeding rate is as variable between different crops as it is between climates and soil conditions. Some of the major factors affecting optimum seeding rates are:

- Available moisture
- Soil conditions and fertility
- Percent germination
- Row spacing
- Planter operation

Some crop species have the ability to compensate for thin seeding or uneven spacing by tillering (such as small grains) or branching (such as soybeans). This allows the crop to fill in gaps so as to make the best use of sunlight. In other crops, such as corn, seed spacing is more important in order to maximize sunlight interception and yield. Particularly in those cases, planters should be adjusted for optimal seed separation as well as population.

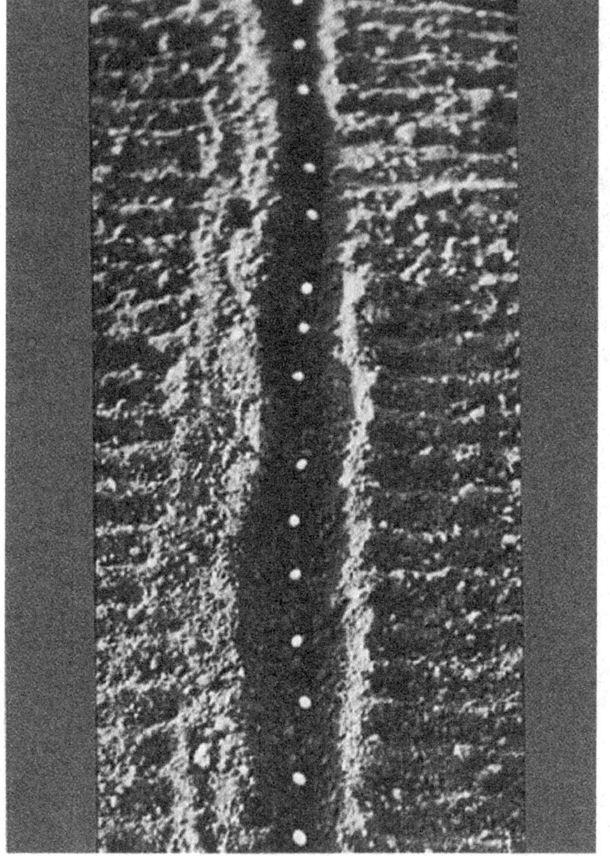

Seeding rates are variable and depend on many factors

MM16633,00025A8 -19-11NOV10-5/5

Planting Machine Capabilities

Field Capacity of Planting Equipment

Field capacity of row-crop planters, grain drills, and broadcast seeders is expressed in acres per hour.

The three factors which concern planting are:

- Theoretical field capacity
- Field efficiency
- Effective field capacity

Let's examine these in detail.

Theoretical Field Capacity

The number of acres (or hectares) that can be planted in one hour if no time were lost is theoretical field capacity (TFC). If a row-crop planter never stopped for refilling, and if it lost no time for adjustments and repairs, or for turning at the ends of the field, the number of acres (or hectares) covered would be equal to the theoretical field capacity (TFC).

The factors determining the theoretical field capacity of a machine are the width of the machine and its speed. Width is usually expressed in feet (ft) and speeds in miles per hour (MPH).

Here is the equation for determining theoretical field capacity (TFC):

$$\text{TFC (Acres/Hour)} = \frac{\text{Width (ft)} \times \text{Speed (mph)}}{8.25}$$

Field Efficiency

The relationship between the number of acres a machine can theoretically cover in one hour (TFC) and the number of acres actually covered (effective field capacity) is referred to as field efficiency (FE).

This relationship is expressed as a percentage of the theoretical field capacity (TFC) of a machine by this formula:

$$\text{FE} = \frac{\text{Effective Field Capacity}}{\text{Theoretical Field Capacity}} \times 100$$

For example, if the row-crop planter planted 89 acres in a 10-hour day, its effective field capacity (EFC) would be 8.9 acres per hour. Differences between the theoretical field capacity (TFC) and the effective field capacity (EFC), 12.73 and 8.9 acres per hour respectively, are due to a number of factors. Among these are time lost when turning at the ends of the field, time lost in making repairs and adjustments, filling seed, fertilizer and insecticide hoppers.

When a farmer gets ready to buy a new planter. he must decide on the size needed. To do this, he must consider the number of acres to be planted and the number of work days available for planting in addition to other economic

Refilling the hopper is just one activity that affects field capacity at planting time

decisions. He must be able to determine what the planter capacity will be. And he needs to know the average acres per hour the machine will plant.

Effective Field Capacity

A machine's capacity is referred to as its effective field capacity (EFC). Field capacities and efficiencies are arrived at through time and motion studies, or simply by observations of how much various machines can accomplish while operating in actual field conditions.

The effective field capacity (EFC) is determined by multiplying the theoretical field capacity (TFC) by the field efficiency (FE) of the machine being considered.

$$\text{EFC} = \frac{\text{Width (ft)} \times \text{Speed (mph)} \times \text{FE (\%)}}{8.25 \times 100}$$

Factors Affecting Field Capacity

There are a number of factors which influence the field efficiency and therefore the effective field capacity of a machine.

Some of the more important factors are:

- Preparing and organizing equipment for minimal downtime
- Size of machine
- Travel speed
- Shape of field
- Condition of equipment
- Condition of soil
- Health and experience of operator

MM16633,00025AA -19-11OCT10-1/1

Introduction

Chemicals are "tools" for a modern farmer. They are as important as mechanical tools for growing successful crops. While chemicals perform many functions, some of the most important for crop production are:

- Improving plant growth
- Protecting against pests and diseases
- Weed control

Chemicals can be used:

- Before planting
- During growth
- After a crop has been harvested

Seed is often treated with chemicals to reduce insect damage and diseases.

Chemicals help ensure good harvest

MM16633,00025AB -19-02DEC10-1/1

Fertilizers

In the past, farmers relied heavily on legumes and livestock manure applications and planned crop rotations to accommodate these nutrient-providing practices. However, gone are the days when most farmers raise livestock.

Today, the nutrient needs of high yielding crops are largely being met by commercial fertilizers. It is estimated that commercial fertilizer is responsible for more than half of the world's crop production.

Compared to natural fertilizer, commercial fertilizers have several advantages.

1. Commercial fertilizers are more concentrated than most natural materials. Less weight and volume are needed to provide equivalent nutrients.

2. The plant food content of commercial fertilizers is known and is uniform throughout.
3. Commercial fertilizers can be tailor-blended to suit a particular requirements.
4. Commercial fertilizers are usually easier to handle than many "natural" materials.

Commercial fertilizers can be applied either as dry ingredients, including most phosphorus and potassium fertilizers, as a liquid such as Urea-Ammonium-Nitrate nitrogen solutions, or a gas, anhydrous ammonia for example. Sometimes they are applied along with pesticide applications or in irrigation water. More information on the role of fertilizers and the different forms is available in chapter 2. Discussion on how the dry, liquid, and gaseous forms are applied will appear later in this chapter.

MM16633,00025E6 -19-02DEC10-1/1

Pesticides

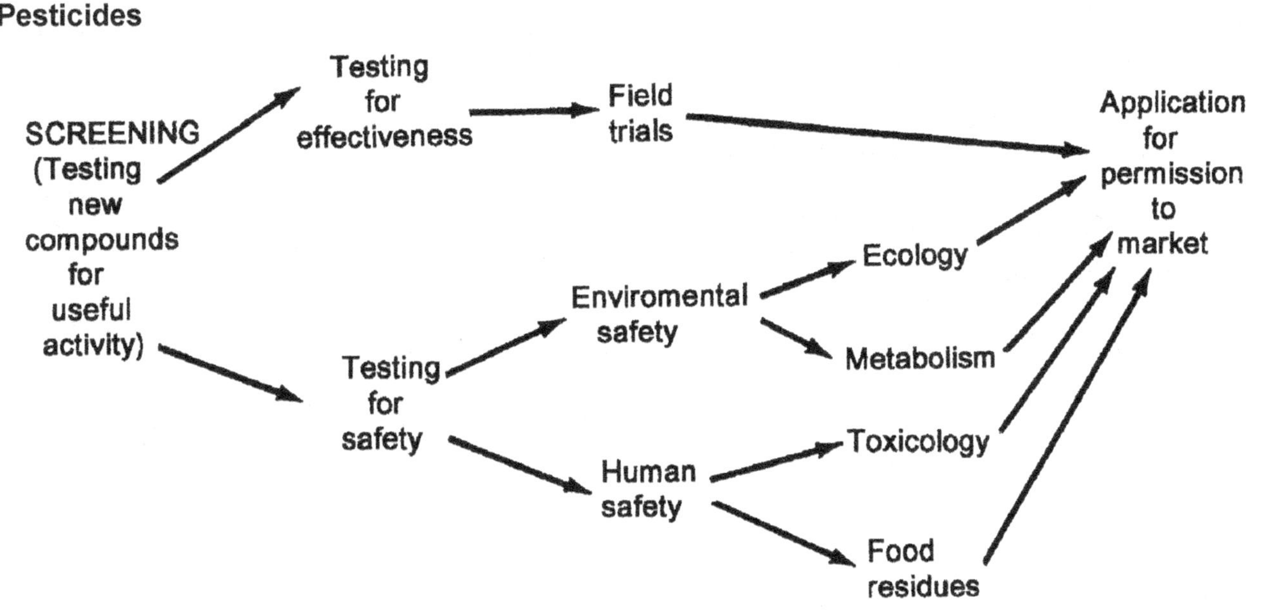

Pesticide development and registration process

Pesticides are natural or synthetic substances used to control pests. A pest is anything that competes, causes injury or disease, or otherwise impedes man, livestock or crops.

The ending "cide" on the word pesticide means killer. Some common pesticides and the pests they are used on

- Insecticides — controls insects
- Herbicides — controls undesirable plants
- Rodenticide — controls rats, mice and other rodents
- Nematicides — controls nematodes
- Fungicides — controls fungus diseases
- Acaricides — controls mites and spiders
- Bactericides — controls bacteria

Regulating Pesticides

The U.S. Environmental Protection Agency (EPA) is the federal agency that registers and regulates all pesticides. Each state also has regulatory powers.

Concerns prompted by chemicals like DDT have resulted in more extensive testing before a pesticide receives a federal (EPA) label. It typically takes six to ten years before a pesticide is registered by the EPA.

Even older pesticides have been carefully examined. In 1988, an amendment to the Federal Insecticide, Fungicide and Rodenticide Act required the EPA to review most pesticides for compliance with regulatory standards. Older pesticides were required to be re-registered.

The high cost of supplying health and safety data needed for reregistration prompted manufacturers to cancel registrations for many pesticides. In the final analysis, modern pesticides have a well documented profile that includes their toxicity to man, animals and wildlife, food safety and impact on the environment.

Pesticide Reporting

Anyone who uses pesticides needs to be aware of requirements for recording and reporting their use. In 1990, for example, California became the first state to require farmers to report all pesticide usage.

The first effort to monitor pesticide use nationwide occurred in 1993. It involved mandatory record-keeping for restricted-use pesticides. A restricted-use pesticide is a chemical that must be applied by a certified applicator due to high toxicity or concerns about the environment.

Information that must be recorded includes:

- Product brand name
- Formulation
- EPA registration number
- Field location
- Crop treated
- Date applied
- Rate per acre
- Acres treated
- Pest(s) controlled
- Total pesticide applied

Since states also regulate pesticides, always check local laws. Regardless of regulations, good record keeping is important and helps avoid serious errors. For example, a pesticide label may restrict what crops are grown the following year. This is because a chemical may leave residues in soil that damage other crops or contaminate them with unapproved pesticide residue. Recording pesticide use for each field prevents costly errors.

MM16633,00025E7 -19-01OCT10-1/1

Ecology and Safety

Despite the value of chemicals, there are valid concerns about the use of pesticides, and fertilizers as well. Pesticides are poisons. Improper application or disposal can adversely affect man, animals, desirable plants, and the environment.

Fertilizers may also have undesirable side effects. Runoff from agricultural land is a major source of surface and groundwater pollution. Nutrients, attached to sediment or dissolved in water, can be carried into lakes and streams. This may result in excessive growth of aquatic plants. When these plants die, the biological breakdown consumes oxygen in the water, leading to "dead" lakes. The lack of oxygen can also cause fish kills.

Dissolved nutrients can also be carried downward through the soil into groundwater. When the concentration of certain nutrient forms exceed critical levels the water may become toxic to fish, animals, or even humans. The nitrate form of nitrogen is the most toxic. It interacts with other blood components and interferes with transportation of oxygen. However, most of the recorded occurrences of nitrate poisoning have been associated with farm wells located too close to manure concentrations rather than runoff or leaching from fields.

The dilemma is that chemicals do so much good and yet may be harmful unless used wisely and carefully.

MM16633,00025E8 -19-01OCT10-1/1

Meeting the Challenge

While concerns over pesticides and fertilizers have made headlines, progress has been made on several fronts. Newer pesticides are generally less toxic and friendlier to the environment than the compounds they replaced. More emphasis is being placed on natural pesticides.

Efforts to reduce chemical use and protect the environment are also taking hold. Four examples are:

- Integrated Pest Management
- Sustainable Agriculture
- Conservation Tillage
- Biotechnology

Integrated Pest Management

Integrated pest management is an approach that tries to strike a balance between chemical and non-chemical control methods.

Integrated Pest Management (IPM) uses all available control methods. This includes cultural and biological controls and the use of resistant varieties. IPM doesn't exclude chemicals, but it considers all alternatives.

IPM was first developed to address insect control, yet it applies to all types of pests. It can be summarized as a concept that integrates all pest control practices into a single program. IPM may not always reduce pesticide use, but it does help ensure that chemicals are used only as needed.

Sustainable Agriculture

Sustainable agriculture is a concept that incorporates IPM. However, it emphasizes alternatives to chemicals. It is also called regenerative farming or low input sustainable agriculture.

There are many definitions of sustainable agriculture. The overall goal, though, is to protect the environment, conserve resources and ensure food safety. Methods used to accomplish those goals are often practices from the past. For example, planting a field to the same crop (continuous cropping) is replaced with crop rotation. This breaks up pest cycles and reduces the need for pesticides. Rotating from row crops like corn or cotton to legumes like alfalfa, also controls erosion. Legumes add nitrogen to soil, so less commercial fertilizer is needed.

Conservation Tillage

When Congress passed the Food Security Act of 1985, it included conservation provisions. Farmers with highly erodible land must use soil conservation practices to be eligible for price supports and other federal programs.

Conservation tillage

No-till, mulch-till and other conservation tillage systems are used to maintain a soil residue cover over at least 30% of the soil. This greatly reduces soil erosion, so less fertilizer and pesticides are deposited in rivers, lakes and other water sources.

Biotechnology

Crop breeders develop varieties which are resistant or compete well with insects and diseases. New biotech tools allow accelerated development of these varieties. In addition to crops that can avoid damage from insect pests, crops have also been developed which allow for use of fewer, better or less toxic herbicides. Increased use of these new genetically-modified organisms (GMO) can reduce the need for pesticide applications.

MM16633,00025E9 -19-08NOV10-1/1

Nonchemical Methods

Many nonchemical methods are used to partly or completely replace pesticides. Weeds, for example, may be controlled by cultivation. Resistant crop varieties ward off diseases and many other pests.

Insect control without chemicals is accomplished with several methods. However, results with nonchemical methods are often variable. Research needed to develop nonchemical controls can be very time consuming. Nonchemical control, though, will grow in importance due to concerns about pesticides.

MM16633,00025F0 -19-01OCT10-1/1

Harvest Aids and Growth Regulators

Other crop chemicals are harvest aids, dessicants and defoliants, and plant growth regulators.

Dessicants

Dessicants are chemicals that dry leaves and other plant parts. They are used to improve harvest efficiency and protect quality of crops such as cotton and legumes and grasses grown for seed.

Defoliants

Defoliants are commonly used in cotton to remove leaves before harvest. This reduces the amount of trash that is harvested and prevents green staining of cotton lint.

Plant Growth Regulators (PGRs)

Plant Growth Regulators (PGRs) are used for several purposes. They may be applied to improve fruit shape or color, to speed crop maturity, or reduce vegetative growth in crops such as cotton. With less vegetation, more sunlight enters the crop canopy. Benefits include better boll retention, less boll rot, earlier maturity, and higher yields.

MM16633,00025EA -19-08DEC10-1/1

Weeds in Agriculture

Worldwide, crop losses from all pests are estimated to exceed $140 billion annually. No one knows for certain the economic losses caused by weeds. However, weeds are a significant factor and may cause losses higher than any other pest. Besides competing with crops, weeds also harbor insects, diseases, and other pests.

Weeds can reduce profits in several ways.

- By reducing crop yields
- By lowering product quality
- By increasing pest control costs
- By interfering with harvest equipment and other machinery
- By harboring Insects and other pests

In irrigated areas, weeds clog water supply and drainage systems. Also, some weed species are highly poisonous. If consumed by livestock, they cause severe illness or even death.

Weeds Compete With Crops for Nutrients, Moisture, and Light

Competition by weeds for the essential elements causes considerable crop loss because nutrients used by weeds are simply not available for the crop.

Weeds use as much water as crops per pound of dry matter produced. Obviously, the more weeds in a field, the higher the water loss.

Photosynthesis requires light, water, and air to take place. Weeds, especially large ones, can shade crop plants and limit the light reaching them. Since light is essential for photosynthesis and growth to take place, shading reduces crop growth. The impact of shading on a crop depends on the degree of shading which also depends on the number and kinds of weeds present.

Highest crop yields are achieved when weeds are controlled

Most weeds are vigorous plants which use large amounts of nutrients. The timing of nutrient uptake, as well as the total quantity used by weeds is important.

For example, pigweed takes up and stores nitrates during the early growth period. These nitrates are not available for the crop to use. This is one reason pigweed is such a successful competitor with most crops. The stored nitrates can also make pigweed hazardous to cattle because of possible nitrate poisoning.

Some weeds have the capacity to obtain water and food directly from crop plants rather than from the soil. An example of such a parasitic weed is dodder. It has no true leaves or roots and is unable to manufacture its own food. After the seeds germinate in the spring, long slender stems wrap around the host plant and smaller suckers penetrate the host to get food and water. If the dodder cannot find a suitable host it dies. Alfalfa and clover are favorite host plants of dodder.

Continued on next page   MM16633,00025AD -19-08NOV10-1/3

Weeds Lower Crop Quality

Weed contamination reduces the harvestability, quality and marketability of many crops. Weeds reduce the nutritional value and palatability of hay and pasture. Consequently, most people are willing to pay more for weed-free hay forages than they would for a weedy crop.

All states have seed laws which may prohibit the sale of agricultural seeds containing certain noxious weed seeds or limit the amount of other kinds of weed seeds that may be present.

Weed seeds in stored grain may retard drying and contribute to mold growth. Small quantities of certain weed seeds such as wild garlic and mustard impart objectionable odors and flavors to flour when the wheat is ground.

Weeds Lower Quality of Animal Products

Cows eating some weeds, such as wild garlic and onion weed, may produce milk with an undesirable flavor.

Weeds Can Poison Animals

Each year, many animals are killed or injured by feeding on poisonous plants. Livestock poisonings are most common during late fall and early spring on overgrazed pasture when grass is sparse and livestock consume plants that would ordinarily be avoided.

Some plants contain naturally occurring toxic substances. Jimsonweed, for example, contains an alkaloid which acts as a hallucinogen. In small quantities, it may not be harmful. However, it can be poisonous if the animal ingests a large enough quantity.

Certain weeds contain photo-active chemicals which travel through the blood stream to all parts of the animal's body. In areas of skin not protected by hair or dark pigmentation, the chemical is activated by light and produces an effect similar to severe sunburn. For example, klamath weed contains a substance which makes the mouths of livestock so tender they cannot eat properly. Spring parsley makes ewes' udders so sensitive that the mothers refuse to let the lambs nurse.

Other plants accumulate toxic substances from the surrounding environment. Lambsquarters, mustard, and pigweed are examples of rapidly growing plants that can accumulate nitrates at levels high enough to be poisonous to livestock. Nitrate concentrations about 1.5% (expressed as KNO_3, dry weight) is generally lethal to livestock. Ruminants are more susceptible than animals with simple stomachs, such as horses.

Other plants can absorb and accumulate toxic minerals, such as selenium, cadmium, copper, lead, and manganese.

Selenium, which is stored in locoweed, accounts for many instances of livestock poisoning.

Weedy crops can cause serious harvesting problems

Weeds Harbor Insects and Diseases of Crop Plants

Some weeds harbor fungal or bacterial disease organisms and insect pests. Black stem rust of wheat spends a part of its cycle on the common European barberry. Besides being a strong competitor for water and nutrients in a corn field, johnsongrass harbors maize dwarf mosaic virus, which attacks corn.

Clogged Channels

Water hyacinth and other aquatic weeds grow in irrigation or drainage ditches, ponds, and lakes. Such weeds reduce the water flow, interfere with drainage, and can require considerable money and effort to keep channels clean. In addition, these plants consume large amounts of water. In arid areas, water used by aquatic weeds is lost and cannot be used for irrigating crops.

Weeds Increase Production Costs

The presence of weeds may require additional cultivations or expenditures for herbicides for control. Weeds also increase machinery wear and slow harvesting. In addition, extra time, labor, and equipment is needed to clean weed seeds from harvested grain.

Weeds May Cause Human Discomfort

There are many weeds that can cause health problems to humans and animals. Some of these problems are:

- Pollen from ragweeds contribute to asthma, hayfever and other respiratory problems
- Skin irritation can result from contact with poison ivy or poison sumac
- Stems, leaves and fruit of some weeds are poisonous, especially to children

Continued on next page

MM16633,00025AD -19-08NOV10-2/3

Changing Weed Problems

Weed problems change as cultural practices, crop selection, and weed control practices are modified. In general, specific weeds are a problem because they thrive and grow under the same conditions being produced for the crop.

The advent of herbicides has changed but not eliminated weed problems. For example, when broad-leaved weeds, which were the major problem in corn, were brought under control by the use of 2,4-D, grasses became a major problem. In a similar way, prickly sida became a serious weed in cotton after trifluralin was used to control other, more aggressive weeds.

The widespread use of herbicides has not led to the development of weed resistance in most cases, but has caused shifts to new weed species which always were tolerant to the herbicide. However, triazine and sulfonylurea herbicide resistant weeds have developed in certain areas of the United States and Canada. More recently, resistance to the widely-used herbicide glyphosate (Roundup™) has emerged in some species, requiring a return to other herbicide options.

MM16633,00025EC -19-01OCT10-1/1

Common Weeds

Before attempting to control weeds, they should be identified. Items to observe include leaf shape, leaf arrangement, type of growth, flower color, and flower shape. There are thousands of weeds. You don't need to know them all. The most practical approach is to become familiar with the weeds common to your area. Some common and serious weeds have a wide range, and appear in many parts of the world. If a weed is found which cannot be identified, contact an Extension agent or an agronomist at your agricultural chemical supplier for assistance.

MM16633,00025ED -19-01OCT10-1/1

Weed Control Methods

Many methods of weed control have been developed. They fall into three groups:

- Mechanical and cultural
- Biological
- Chemical

Mechanical and Cultural Weed Control

Although herbicides have become more important in weed control, mechanical and cultural methods are still widely used.

Tillage — This is a common weed control method, but has become less common due to an emphasis on reduced tillage along with effective herbicide options for killing weeds. However, a resurgence in tillage and cultivation for weed control is occurring as increasing production of organic crops has eliminated chemical options for those growers. It is also used to incorporate herbicides and loosen soil to improve aeration and water penetration. Tillage may be done before planting and after crop emergence. Implements such as rotary hoes, spike tooth harrows, and finger weeders are used after row crops emerge.

Row crop cultivation has decreased since the advent of effective chemical weed controls. On soils in good tilth, weed control is often the main benefit of row crop cultivating. Improperly adjusted cultivators can damage a crop.

The rotary hoe can be very effective for early control soon after the weed seeds have germinated and before or soon after the weed seedlings emerge. Even where a herbicide has been used, the rotary hoe can be very effective, especially if delayed rainfall decreases effectiveness of the herbicide. The rotary hoe can help to control the first flush of weeds and may even give very slight incorporation of the herbicide to increase effectiveness for the next flush of weeds.

Hand pulling or hoeing — This is probably one of the oldest methods of controlling weeds. Large scale use is limited, but it is still widely used in home gardening and developing countries. It is a practical way to remove a few scattered weeds in a field to prevent seed production, particularly of noxious or hard-to-control weeds. Pulling is most effective on annual or biennial weeds. It is less useful for perennials, because they grow back from root parts left in the soil.

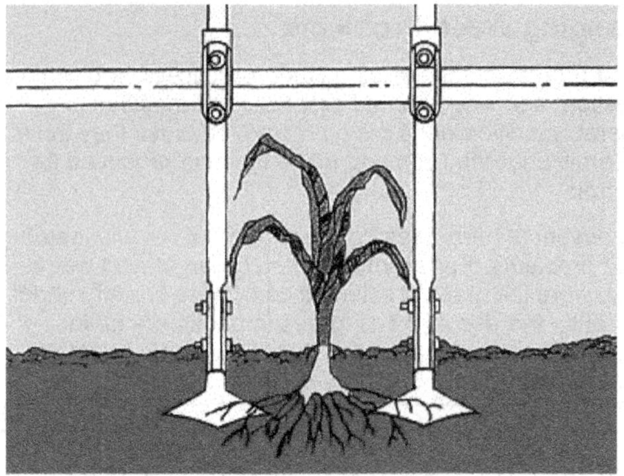

Cultivating too close or too deep cuts crop roots

Mowing and grazing — Weeds can be mowed to prevent seed production and to remove unsightly growth. To be most effective, mowing must be close to the ground to ensure that all branches, tillers, or stems are cut. Close mowing, however, is not always possible and weeds often regrow quickly. Grazing has sometimes been used to control weeds in pastures, and along ditches, fence rows, and roadsides. Sheep and geese are used for weeding some crops.

Flooding — Flooding controls weeds by excluding air that is essential for plant growth. It is only effective if the entire plant is covered for a fairly long period of time. It is used in conjunction with crops like rice that can grow in the flooded area.

Burning — Burning is sometimes used to control weeds. Burning rangelands removes plant cover, including weeds. But many seeds remain viable in the soil and valuable organic matter is destroyed. The practice is being restricted in some areas due to air quality concerns.

Selective weed killing in a crop field is possible with specialized flaming equipment. Temperatures as high as 2,000°F (1,090°C) can be reached by burning fuel oil or propane. Flaming is rather expensive, but effective when used in cotton. Smaller, more tender weeds are destroyed by the heat, but the woody cotton stems are not harmed significantly by the flame.

Flaming is used on a very limited basis. The cost of fuel makes it less economical than other weed control methods.

Continued on next page   MM16633,00025EE -19-01OCT10-1/2

Smothering — Weeds can often be smothered by competitive crops or by various kinds of mulches. Planting patterns that quickly produce a complete ground cover enable some crops to compete effectively by shading weeds.

Mulches are often used to control weeds around ornamental plants or in high-value crops such as strawberries and melons. Plastic, paper, straw, wood chips, and manure are just a few of the materials used.

Biological Control

Biological weed control involves the use of living organisms such as insects or fungi that attack certain weed species. Biological control is particularly well suited to situations where other control methods, such as an effective and economical herbicide, are not available.

Chemical Control

Although herbicides are sometimes used to eliminate tillage, the more common procedure is to combine chemicals, tillage, and other cultural practices in a carefully designed, well coordinated weed control plan. When compared to other control methods, herbicides have the following advantages:

- Appropriate herbicides may be applied to control weeds in crops where row spacing is too narrow for cultivation
- Pre-emergence application of herbicides provides early season weed control, allowing the crop to grow without competition during the critical early part of the growing season

Smothering can help control weeds

- Selective herbicides reduce the need for cultivation, which can injure crop roots as well as weeds
- Herbicides help maintain soil structure by reducing the need for tillage
- Erosion can frequently be reduced by vegetation or cover of crop residue while herbicides are used for weed control
- Many weeds, especially perennials which cannot be effectively controlled by other means, are susceptible to herbicides
- Although herbicides require significant amounts of energy to produce, transport, and apply, they can reduce the total amount of energy required for weed control when compared to the mechanical approach

MM16633,00025EE -19-01OCT10-2/2

Disease Control

Plant diseases are among the greatest hazards in man's continuing struggle to produce food. Many diseases can spread explosively and cause wholesale destruction of crops. Because of the heavy damage they can inflict, diseases often limit the kinds or varieties of crops that can be produced in a particular region.

Pathogens

The disease-producing agent is called a pathogen. There are three important groups of biological pathogens.

- Fungi
- Bacteria
- Viruses

Fungi As Plant Pathogens

Fungus diseases are often harder to control with chemicals than insects. The fungus lives in close contact with its host plant. Fungi often reproduce rapidly and new plant growth is subject to infection. Multiple fungicide applications may be needed. This is especially true with vegetable and fruit crops since appearance is important.

Seed and soil-borne diseases — In many field crops the most destructive diseases are often caused by fungi on seed or in the soil. Treating seed with fungicides reduces seed-borne diseases. Seed treatment also protects against soil-borne diseases Pythium, fusarium, and rhizoctonia are soil borne organisms that affect many crops.

They cause seed rots, seedling blights, and damping-off. The latter is a common disease of seeds and seedlings. It may prevent seeds from sprouting, seedlings from emerging, or cause stem rot of seedlings, causing them to fall over.

As with other types of diseases, chemicals are not the complete answer to preventing seed and seedling diseases. It requires a total program that includes:

- Planting high-quality seed
- Planting at recommended soil temperature
- Delaying planting if soil is too wet
- Planting resistant varieties if available

Bacteria As Plant Pathogens

Compared to fungi, relatively few major plant diseases are caused by bacteria. However, a number of bacterial diseases can be highly destructive. At one time, for example, soybeans were considered a "disease-free" crop. Soybeans have now been grown in the U.S. long enough to experience severe disease damage, including bacterial pustule and bacterial blight.

Bacterial blight is one of the most widespread soybean diseases in the Midwest. The bacteria are seed-borne and the disease can be avoided by not planting seed from infected fields.

Blight is also a serious disease of cotton in areas such as the Mississippi Delta, where warm wet weather is common during the growing season. It reduces yield through defoliation (loss of leaves), boll infection, and spotting of cotton lint.

Symptom	Example Disease
Yellowing	Bacterial Wilt of Alfalfa
Wilting	Southern Bacterial Wilt of Tomatoes, Potatoes and Tobacco Bacterial Wilt of cucumbers, melons, squash, and pumpkins
Blight	Fire Blight of Apples and Pears Wildfire of Tobacco Bacterial Leaf Blight of Rice
Spot	Bacterial Spot of Stone Fruits Angular Leaf Spot of Cotton
Canker	Fire Blight Bacterial Canker of Stone Fruit
Rot	Soft Rot of Vegetables
Gall	Crown Gall of Ornamentals Leafy Gall
Scab	Scab Potato

Table 1 — Bacterial Disease Symptoms

Table 1 — Bacterial Disease Symptoms

Types of Bacterial Diseases — The symptoms produced in plants by bacterial diseases are similar to those produced by other pathogens. Each pathogen usually produces a definite pattern of symptoms in the host plant. Some of the bacterial disease symptoms are shown in Table 1.

Spreading Bacterial Diseases — Plant-infecting bacteria are not spore formers and cannot remain alive very long away from a suitable host. But they can survive longer in water and are mainly dependent upon wind-driven rain for spreading from plant to plant. Other means of spreading bacterial diseases include:

- Dust storms
- Irrigation water
- Machinery during cultivation and harvest
- Insects

Viruses As Plant Pathogens

Viruses are submicroscopic entities that multiply and survive only inside living cells. Few plants and animals are immune from viruses.

Few chemicals are available to control virus diseases. In most cases planting resistant varieties and use of cultural practices is the only available defense.

Viruses are spread by certain sucking and chewing insects. Therefore, the threat of disease infections is highest in years when there are high populations of insects that transmit viruses. Mosaic diseases of corn and soybeans, for example, may erupt when aphid populations are high.

Bud blight of soybeans is caused by the tobacco ringspot virus. While it has appeared sporadically, it occasionally caused losses of 25% or more in these crops.

Continued on next page

Continued on next page

MM16633,00025AE -19-08NOV10-1/2

Virus Symptoms — Viruses can induce a wide variety of symptoms in host plants. Some, such as spotted wilt or curly top of tomatoes, can kill the plant. More commonly, the result is lowered product quality and reduced yields. There are two clearly defined groups of symptoms:

- Mosaics cause mottling and chlorotic spotting of leaves
- Yellows which cause yellowing, leaf curling, dwarfing or excessive branching

Virus Transmission — Some viruses are transmitted in infected seed, but only a small percentage of the seed from an infected plant will usually carry the virus.

Insects are an important natural method of spreading viruses. Insects that vector (transmit) diseases include aphids, thrips, whiteflies, leafhoppers, and beetles. Aphids generally vector mosaic-type diseases, and leafhoppers transmit "yellow" type diseases.

Cultural practices and timely insect control are key ways to deal with virus diseases. Select resistant varieties when available. If possible, delay planting until after insect activity ceases. In the eastern U.S., for example, barley-yellow dwarf is avoided by planting when colder weather halts aphid activity. This occurs after the Hessian fly-free date.

Good weed control also aids efforts to avoid virus diseases. Weeds harbor insects and diseases. Controlling weeds in and around fields eliminates host plants that aphids and other insects collect in before attacking crop plants.

General Approaches To Disease Control

Satisfactory control of most plant diseases requires the application of several control measures and usually involves an integrated program of environmental, biological and chemical factors.

As mentioned earlier, correct diagnosis is essential to plant disease control. Then, control involves the application of one or more of the following principles:

- Avoidance — avoiding disease by planting when, and or where, pathogens are ineffective or absent
- Exclusion — keeping pathogens out of a "clean" area
- Eradication — eliminating the pathogen source, whether an infected plant, field or region
- Protection — preventing an infection by using a chemical or physical barrier to keep pathogens out
- Resistance — using plants that tolerate, resist or are immune to the disease
- Therapy — reducing the severity of disease in an infected plant

The first five of the principles are methods of disease prevention. The sixth represents the cure. At present, plant disease control measures are predominantly preventive. From a farmer's viewpoint, the general approaches to disease control are reduced to three practical methods:

- Cultural and biological control
- Control through disease resistance
- Chemical control

MM16633,00025AE -19-08NOV10-2/2

Insect Control

Although insects have many powerful survival abilities, there are ways to control them. The primary control is to prevent them from spreading into new areas. If an insect infestation has already been established, first decide what, if any, control measures should be used. This decision is usually dictated by economics. Sound Integrated Pest Management (IPM) principles dictate that control measures are usually applied only if the expected losses without control exceed the cost of the control. Insect problems are both complex and diverse. Large concentrations of host plants in a crop field or worse, a growing region, help to create and aggravate insect problems. There are no simple solutions. To cope with insect problems effectively, farmers should:

- Correctly identify the insect causing the damage
- Know, or become familiar with, the life cycle of the insect, particularly the times in the cycle when it is most susceptible to control
- Apply effective and economical control procedures
- Prevent re-infestation

There are four methods of insect control:

- Natural
- Biological
- Cultural
- Chemical

Natural Control

Natural control is the reduction of insect populations by the forces of nature. Natural controls include climate, topography, and natural enemies.

Climate — Weather conditions, especially temperature, affect insect activity and rate of reproduction. Climate affects insects directly and indirectly by influencing the growth and development of their host plants.

Direct killing may result if the weather acts against the insect at any stage of its life cycle. Some insects are killed by low winter temperatures. Insects with one or more stages that live in water are controlled if free water is not available, because low rainfall or high temperature cause drying. Direct killing is by no means a certain process. For example, winter-killing is moderated by available protection and the inherent cold resistance of the hibernating stage. Insects seek shelter for overwintering. They burrow under bark, in plant debris, or underground. Snow also protects the insects by insulating them against extremely low temperatures.

The seasonal population of plant-eating insects is closely related to growth of the host plants. Unusual weather conditions can change the normal pattern so that increased or decreased damage results. For example, if a cool spring retards the cherry fruit fly, less damage results because fewer insects are present to infest the fruit before harvest.

Topography — Features such as mountains and large bodies of water restrict the spread of many insects. Major topographical features which affect the climate of an area also limit the distribution of insect species.

Other features of the landscape can have similar effects. Soil type is a prime factor affecting wireworms. Some species live in heavy, poorly-drained soil, others in light sandy soils. Soil type also affects plant distribution, which in turn affects the insect population.

Natural Enemies — Birds, reptiles, fish, diseases, and other predatory insects are natural enemies which help control insect populations. Of the natural enemies, predatory and parasitic insects are the most important. Of all insect species, more than half feed on other insects, some of which are crop pests.

The interrelationships between pests and their enemies are complicated. In general, a population increase by pests is followed by an increase in predators. As the pests increase, the predators can find them more easily. As a result, the populations of predators increase. Eventually, there are so many predators that pest numbers decline to the point that the predators are themselves restricted because of a lack of feed. However, pest populations can often reach epidemic numbers before there are enough enemies to achieve worthwhile control.

A good example of a predatory insect is the ladybug or ladybird beetle, which feeds on aphids, scales, and other soft-bodied insects including the soybean aphid, greenbug of wheat and sorghum, and the corn borer. Adult beetles eat the aphids and lay eggs on the infested plants. When the beetle larvae hatch, the aphids are a handy food supply. Each larva can eat up to 500 aphids (300 per day or 60 per hour) before pupating (entering the next stage of metamorphosis).

Biological Control

Biological control of pests, such as by predators, may also be achieved by insect or disease organisms that have been introduced by man.

These biological control programs involve searching for a parasite or predator for a particular insect pest. Since most major insect pests were introduced to the U.S. from foreign countries, the searches often take place in the source country. Most projects are conducted by governmental agencies.

An example of biological control is using laboratory-reared wasps to control cereal leaf beetles. The wasps lay eggs in the living beetles or in the beetle eggs. When the wasp hatches, the pest is killed.

The sterile-male technique and use of baits and repellents are other approaches to biological control.

Cultural Control

Cultural controls may be used when other methods are not available to combat an injurious species and to supplement other controls. Cultural methods can be economical and are particularly suited to pests of low-value crops.

Cultural controls include:

Continued on next page

MM16633,00025AF -19-09NOV10-1/4

- Crop rotation
- Trap crops
- Tillage
- Timing of operations
- Resistant varieties
- Mechanical controls

Crop Rotations — Continuous production of a single crop species on a piece of land often encourages pests to build up. Changing crops to a species which is not a host usually aids in control of these pests. Crop rotations are most effective against insects which have long life cycles and infest the same crop during all stages of growth. Changing crops cuts such insects off from their primary or only food supply.

Trap Crops — Small plots of a host plant the insect favors, located near susceptible crops, are called trap crops. After the insect pests have been attracted to the "trap," they can be killed with an insecticide.

Tillage — Tillage operations can help control insects by changing the physical conditions of the soil, mechanically interfering with some stage of the insect's life cycle, removing host plants, or increasing the growth and vigor of plants to improve their resistance.

Tillage is often involved in residue-management programs. For example, shredding and plowing-under corn stalks is an important part of a control program for European corn borers.

Shredding stalks helps eradicate plant-borne insects. Rotary-cutting and chopping action can control insects such as the European corn borer, and pink bollworm in cotton.

Insects suffer greater winter-kill if the stalks are pulverized. In some states, cotton stalks must be pulverized and plowed down by a certain date. The date is set by law in a pink-bollworm control program. Plowing at least 6 inches deep immediately after shredding can reduce pink bollworm by 95%. A flail chopper usually kills more larvae than a rotary cutter because of greater shredding and pounding of the debris.

Timing — Proper timing of planting or harvesting operations can be used to control insect damage if the host plant is susceptible only for a brief period or if the infesting stage of the insect's life cycle is short.

For example, prevention of Hessian fly damage in wheat is most easily achieved by delaying planting until cooler temperatures prevent further fly reproduction.

At the other end of the season, crops should not be left in the field after growth is completed if they are susceptible to pest attack. For example, wireworm damage to mature potatoes causes a serious quality reduction. Since damage increases if the crop is left in the soil, harvesting should begin as soon as maturity is reached.

Resistant Varieties — Resistant strains of many crop plants are available. There are three ways that insect

Plowing down corn stalks exposes European corn borers to winter kill

resistance can be incorporated into a particular crop variety:

- Including genetic changes which do not allow the insect to mature
- Making the variety more vigorous so it is better able to resist insect attacks
- Including characteristics that make the plant less appetizing to the insect

One of the earliest uses of gene-transfer technology in agriculture was to introduce a gene from Bt (Bacillus thuringiensis) into crops. Some strains of Bt, which is a naturally-occurring soil bacterium, produce proteins that kill certain insects. When the insects ingest the protein produced by Bt, the function of their digestive systems is disrupted, producing slow growth and, ultimately, death. Today, Bt is a common source of resistance for insect pests in corn, cotton, and potato, and other Bt crops such as rice are soon to follow.

Mechanical Controls — Devices that affect insects directly or seriously alter their physical environment are called mechanical controls. They are different from cultural controls in that the control is achieved by physical measures rather than farming practices. Screens, barriers, traps, electricity, X-rays, heaters, and refrigeration equipment are examples of physical or mechanical control devices. Both heat and cold are used to reduce insect populations in grain elevators. No insect can survive temperatures above 130° to 140°F very long (54° to 60°C).

Chemical Control

Chemical control is the reduction of insect populations by the use of chemicals that:

- Poison the insects
- Repel the insects from specific areas.
- Attract insects to a place where they can be killed by other chemicals or some other method.

Continued on next page

MM16633,00025AF -19-09NOV10-2/4

Chemicals may be applied to seeds, plants. or soil.
Effectiveness depends on the characteristics of the
chemical, plus the timing and method of application.

MM16633,00025AF -19-09NOV10-3/4

Nematodes

Nematodes are eel-like animals. Most are invisible. A
common characteristic is a hollow tube used to puncture
plant cells and withdraw nutrients. Plant symptoms often
mimic problems such as a nutrient deficiency or moisture
stress. Positive identification often requires laboratory
analysis of soil samples. There are more than 10,000
nematode species, but only a few damage crops. They
can be divided into two major groups that feed on above
and below-ground plant parts.

- Endoparasites — enter into and feed inside plant tissues
- Ectoparasites — remain outside the plant and feed on
  surface cells.

Most nematodes feed on underground plant parts and
attack small roots, although tubers, corms and similar
organs may also be affected. Root feeding results in
destruction of roots and impaired uptake of water and
minerals.

Root-knot nematodes cause important plant diseases in
warm climates and in greenhouses. The saliva which is
injected into the plant cells causes formation of swollen
and distorted roots.

Cyst-forming nematodes cause infected cells to enlarge,
although not as much as the root knot group. At death,
the outer covering of the female cysts becomes tough and
hardened. The covering, which is resistent to chemicals
and other microorganisms, provides protection in the
soil for several hundred eggs for as long as 10 years,
increasing problems in nematode control and crop rotation.

Other nematodes cause damage, including:

- Root lesions
- Root elongation
- Swollen root tips
- Reduced or "bushy" root growth
- Crinkled and distorted stems or bulbs
- Leaf sons

Because nematodes are so small and attack roots, their
presence and impact on yield is often hidden. However,
their feeding results in significant yield loss in soybean,
cotton, and corn. The best way to identify the presence
of nematodes is to submit soil tests to diagnostic labs.
Treatments include resistant varieties, soil insecticides,
and seed treatments. Crop rotations can also be
effective in reducing some nematode populations. Certain
nematodes are known to transmit viruses. Also, nematode
attacks can increase the chances of a plant being invaded
by fungi and bacteria that cause root diseases, such as
fusarium wilt of cotton.

MM16633,00025F1 -19-09NOV10-1/1

Insecticides

Insecticides are poisons that affect specific parts or organs of an insect. For an insecticide to be useful, it must kill the insect or break its life cycle without destroying or injuring valuable plants, animals or beneficial insects. Similar insecticides may be grouped in various ways. The "point of entry" classification has three groups.

- Stomach poisons are sprayed or dusted on a plant. The insect swallows the poison when it feeds on the plant. Stomach poisons are suitable for chewing insects such as codling moth, cotton bollworm, gypsy moth, and cabbage looper. Stomach poisons are "protective" — they can be applied to the plant preceding the insect infestation.
- Contact poisons kill when they come into contact with the insect. These insecticides kill the insect they reach during application, but usually are less effective on insects that arrive later.
- Systemic poisons are absorbed into the sap of plants through the roots or foliage. The material then kills insects which feed on the plant. Some systemic insecticides are also contact poisons.

Insecticides are more frequently grouped by their chemical characteristics. The major divisions are:

- Inorganics
- Organics

Inorganic Insecticides

Inorganic pesticides are materials that do not contain carbon. They are often crystalline or salt like in appearance. Most inorganics have been replaced by more efficient synthetic compounds. Two inorganic insecticides still being used are cryolite and sulfur. Sulfur is used to protect vegetables and grapes. Sulfur dusts and sprays are used to control mites. Sulfur is also a fungicide.

Organic Insecticides

Organic insecticides contain carbon. They may be natural insecticides derived from plants (botanicals), or they may be synthetic compounds resulting from organic chemistry. Botanicals are common household insecticides. But, high cost and a limited residual life limit their use in agriculture. The synthetic organics dominate the field. The three most widely used groups are:

- Organophosphorous
- Pyrethroids
- Carbamates

Organophosphorous — They are also called phosphates, organophosphates or O.P. insecticides. Most are contact or stomach poisons. Some have high vapor pressure and kill by fumigating action. Several are systemic.

The organophosphorous insecticides are cholinesterase inhibitors or nerve poisons. They affect the nervous system of target insects and kill rapidly. They break down quickly, so they're desirable from an environmental standpoint. However, many are highly toxic to man and animals. An exception is malathion, a widely used D.P. insecticide. It is a common ingredient in flea powders for pests and other domestic animals.

Pyrethroids — The pyrethroids are synthetic versions of pyrethrum, a botanical insecticide. Unlike pyrethrum, pyrethroids are stable and have an effective residual period. But, they share common traits with pyrethrum. Pyrethroids control a wide spectrum of insects. Most are relatively low in mammalian toxicity. They are contact and stomach poisons. They are biodegradeable and are applied at low use rates: generally less than one-tenth pound of active ingredient per acre.

Carbamates — This group is derived from carbamic acid. Like the phosphate insecticides, carbamates are cholinesterase inhibitors. Some kill on contact while others are systemic. Mammalian toxicity runs from low to extremely high. Carbaryl, a widely-used carbamate is non-systemic and low in toxicity. Aldicarb is systemic. It has one of the highest oral toxicities of any insecticide active ingredient.

Other Insecticide Groups

Several other types of insecticides play an important role in various insect control programs. Following is a brief description of these insecticide groups.

Microbials — Bacteria, viruses, fungi and other naturally occurring microorganisms cause diseases fatal to insects. A widely-used microbial insecticide is the bacterium Bacilfus thuringiensis (Bt). When insect larvae (caterpillars) ingest the active ingredient, it destroys their gut. Insects stop feeding soon after eating the bacterium and die within several days.

Several Bt strains are sold as insecticides. They are nontoxic to beneficial insects, making them valuable for integrated pest management programs.

Insect Growth Regulators — This relatively new group of chemicals affects insect growth in a way that reduces populations. Diflubenzuron was the first insect growth regulator (IGR) registered for field crops. It controls insects like the cotton boll weevil by preventing immature stages from producing chitin (pronounced kytin). Chitin is a hard outer covering necessary for survival.

Azadiracthin is an IGR extracted from seeds of the neem tree. It prevents insects from shedding their outer covering (cuticle). This process is known as molting. Insects controlled include whitefly, armyworm, and cabbage looper.

Spray Oils — This group is made from crude petroleum and refined and formulated as insecticides. Oils kill insects and eggs by smothering them. There is no residual activity and complete spray coverage is needed.

Dormant oils are applied to tree fruit and nut crops after leaves drop. They contain impurities that can injure tender foliage. Foliar or refined oils can be used when foliage is present.

Continued on next page

MM16633,00025EF -19-21OCT10-1/2

102615

PN=69

Emulsion or flowable oils contain about 20% water. Water evaporates after application. Emulsive oils contain an emulsifying agent and no water. If water is added in the spray tank, agitation must be provided to form an emulsion.

Synergism

Some chemicals have the ability to greatly increase the killing power of certain insecticides. If the toxicity of the combination is greater than the sum of the two chemicals used separately, it is called a synergistic action. Synergism is somewhat like adding 2 plus 2 and getting 6 or 9.

The synergist may be inactive when used alone, or may be insecticidal. When applied together pairs of organophosphorus insecticides frequently have a synergistic effect. Synergism is important for:

- Increased insecticide effectiveness
- More economical control because less material is needed

Synergism can also produce a greater than anticipated hazard to people working with the chemicals. Extra caution should be exercised if chemicals are combined.

Insect Monitoring

Responsible pesticide use not only means knowing how to spray, but also when to spray. University entomologists have established economic threshold levels for many insects. If an insect population is at or above the threshold, insecticide use is justified. One or more methods are used to track insect populations. They are:

- Visual inspections
- Use of sweep nets
- Pheromone baited traps

Factors like stage of crop growth, insects present, climate and cultural practices determine scouting frequency. Conservation tillage is an example of cultural practices that require more monitoring. Decaying vegetation and cool, moist conditions in spring attract certain insects, including pests that attack seeds and seedlings. Thus, when changing cultural practices, be aware of possible changes in insect pressure to develop a sound control strategy.

Insecticide Hazards

Everyone who uses insecticides must be aware of the dangers involved. Always read the product label before use and carefully follow all directions and precautions. Be sure you know the safely measures for proper handling, application and storage of insecticides. Become familiar with posting requirements for hazardous materials. Posting fields involves placing warning signs to alert anyone who comes into the area that a field has been treated with a toxic chemical.

Also, take time to learn about possible environmental hazards of insecticides. For example, some compounds are highly toxic to fish and/or bees.

Accidental Poisoning

Nearly all pesticide accidents result from failure to follow recommended procedures for handling, application, and storage.

Pesticide poisoning can result from three kinds of exposure:

- Oral — through the mouth and digestive tract
- Dermal — through the skin
- Respiratory — through the lungs and respiratory system

There is a wide difference in the toxicity of various insecticides.

Avoiding Illegal Residue

Insecticide residue on crops has the potential for oral poisoning. Insecticides, like other pesticides, are subject to regulations governing a specific time interval called the pre-harvest interval, or PHI, between the last allowable application and harvest. This information is clearly stated on the label. Failure to adhere to PHI restrictions can result in severe penalties. Always check product labels for the time interval between application and harvest before applying insecticides and all other pesticides.

⚠ **CAUTION: Insecticides may enter the body through skin or lungs, not only during application, but also from contact with treated plants after application. Definite waiting periods are specified for chemicals before anyone can safely enter the field following application.**

Labels

Pesticide labels are important to every pesticide user. The information and instructions on the label are developed at great expense during years of research by the chemical manufacturer. The label tells how to use the material correctly and safely.

Read the label every time you use the material. Don't rely on your memory.

Label formats are similar for all pesticides.

Development of Resistance

The insect population in a given crop is made up of individuals with varying resistance to chemicals. If repeated applications of the same insecticide or class are made, the most resistant individuals will be the ones that survive and reproduce. Eventually, a resistant population may develop. This process is most common in crops that receive routine insecticide applications. Thus "insurance" spraying may eventually produce resistant populations of the insect species the insecticide initially controlled. Strategies to delay the development of resistance are to spray only when necessary and alternate the use of pesticides with other insecticides that are effective on the target pest.

MM16633,00025EF -19-21OCT10-2/2

Application of Liquid Chemicals

Types of Sprayers

More pesticides are applied with sprayers than any other kind of equipment. Liquid fertilizers can be applied in the same equipment, and are often mixed with the pesticides. There are many types and sizes of spray equipment. Depending upon the equipment used, and recommendations for the product and liquid carrier (i.e. water or liquid fertilizer), applications may be made at pressures ranging from near zero to 1,000 psi (6,900 kPa). Application rates vary from a few ounces (milliliters) to several hundred gallons per acre (liters per hectare). In size, the equipment may be as small as an aerosol can or as large as a helicopter or airplane.

Although there are many variations and combinations of types, most liquid-pesticide-application equipment falls into five categories.

- Hand-operated sprayers (5–40 psi, 30–275 kPa)
- Low-pressure sprayers (10–150 psi, 70–1,000 kPA)
- High-pressure sprayers (to 1,000 psi, 6,900 kPa)
- Air carrier sprayers (to 1,000 psi, 6,900 kPa)
- Foggers and rope wick applicators

Each type of equipment has distinctive features.

Hand-Operated Sprayers

Hand-operated sprayers are commonly used by home gardeners and others whose pest problems are relatively small. However, some farmers and custom operators also find hand-operated units useful for:

- Spot treatments in a large area
- Small jobs where field-sized equipment is not required
- Treatment of hard-to-reach areas inaccessible to large powered equipment

Compressed Air Sprayers — These sprayers are easy to operate, and relatively inexpensive to buy and maintain. The capacity of the tank usually ranges from one to 5 gallons (3.8 to 19 liters). Pressure is provided by a hand-operated air pump which fits into the top of the tank. Air compressed in the tank above the spray material forces the liquid out of the tank through a tube. A valve at the end of a short length of hose controls the flow of liquid. Agitation is provided by shaking the tank. Normal operating pressure is between 5 and 40 psi (30 and 275 kPa) and is maintained by occasional pumping.

Another version of the compressed-air sprayer uses a pre-charged cylinder of air or carbon dioxide to provide pressure. These units include a pressure-regulating valve to maintain uniform spray pressure. Wheel mounting provides easy portability. Pesticides may be applied through a hand gun or a short boom.

Knapsack Sprayers — These sprayers have a small piston or diaphragm pump. The level operated type, which is powered by the operator, is widely used In all parts of the world. Some newer models are powered by a small gasoline engine. An air chamber helps smooth out pump pulsations and maintain uniform pressure. Tank capacity ranges from 2 to 6 gallons (7.5 to 23 liters) and pressures up to 180 psi (1,240 kPa) can be developed.

Agitation of spray material in the tank is provided by a mechanical agitator or by bypassing part of the pumped flow back to the tank.

As the name implies, a knapsack sprayer is carried on the operator's back, It is also called a backpack sprayer. The hose and nozzle are similar to those used on compressed air sprayers. A knapsack sprayer lets the operator conveniently treat small patches of weeds in crops and around field borders. The knapsack sprayer is also used in small research plots.

Low-Pressure Sprayers

Low-pressure sprayers are the most widely used type of field application equipment. They are relatively inexpensive and are adapted to many uses including pre or post emergence application of chemicals to control weeds, insects, and diseases. They can also be used for other jobs such as spraying cattle. Low-pressure sprayers are available in many models and types and may be mounted on:

- Tractors
- Self-propelled sprayers
- Trailers
- Trucks
- Aircraft

Low-pressure units usually operate in the 20–50 psi (140–345 kPa) range and apply from 5 to 60 gallons per acre (gpa) (50–560 liters per hectare).

Controlled Droplet Applicators — A controlled droplet applicator (CDA) applies low volumes of pesticide mixtures. Typical volume is one to three gallons of spray mix per acre. They are most often used with foliar-applied herbicides.

Instead of nozzles, CDAs have a spinning disc or cup with serrated edges onto which pesticides drop. Cups or discs spin at speeds of 1,000 to 6,000 rpm, producing small droplets that are relatively uniform in size. They are powered by electric or hydraulic motors. CDA units range from hand held or backpack units to larger models that are mounted to a boom and attached to a tractor.

NOTE: Use low volumes of water only if recommended on the product label. The advantage of using low gallonage is that it minimizes time spent refilling spray tanks.

Continued on next page

MM16633,00025B0 -19-08DEC10-1/3

Tractor-Mounted Sprayers — These sprayers usually hold from 150 to 500 gallons (570–1,890 L). The tank can be mounted in one of several positions on the tractor. The pump is usually attached directly to the PTO shaft but may be driven by a hydraulic motor or other means. Booms may be mounted in the front, rear, or "belly" positions. Broadcast applications may also be made with a nozzle cluster. Tractor-mounted units can be combined with other equipment such as planters, cultivators, or tillage implements.

Self-Propelled Sprayers — Also know as high-clearance sprayers have evolved from tractor-mounted sprayers. They have a frame high enough to clear corn, cotton, tobacco, and other tall crops. The spray boom may be raised or lowered, depending on crop height and application requirements.

Trailer-Mounted Sprayers — These sprayers are built on wheels and towed through the field by a tractor. Tank capacity ranges up to 1,000 gallons (3,785 L). The pump is mounted on the tractor or sprayer and driven by the tractor PTO shaft or a hydraulic motor. Boom lengths vary from 12 to 60 feet (3.7 to 20 meters) or more. Trailer sprayers are available in high clearance models.

Skid-Mounted Sprayers — Skid-mounted sprayers may be placed in a pickup or flat-bed truck. Pump power is supplied by an auxiliary engine. Larger truck-mounted units can be equipped with flotation tires so they can operate under relatively wet conditions. They are best suited for large acreages or custom operation. Such units have tanks holding up to 2,500 gallons (9,465 l) and booms measuring up to 80 feet (26 m).

Aircraft-Mounted Sprayers — Principal advantages of aircraft sprayers compared to ground equipment are:

- Rapid ground coverage
- Application can be made in places and at times when ground equipment cannot operate

Unless properly calibrated and operated, aircraft sprayers are not as thorough in applying material as ground rigs, especially in tall crops. However, coverage is adequate for most needs. Both fixed-wing airplanes and helicopters are widely used.

Helicopters are more maneuverable and are not restricted to operating from a landing strip. But, they are more costly than fixed-wing units and have a smaller payload.

Because of the limited carrying capacity of all types of aircraft, spray materials are usually applied in a more concentrated form than for ground application. Volumes of 3 to 15 gallons per acre (28 to 140 L/ha) are common. Application is made at a height of 3 to 25 feet (0.9 to 7.6 meters) above the tops of the plants at 80 to 140 mph (130 to 225 km/h).

High-Pressure Sprayers

High-pressure sprayers are similar to low-pressure sprayers. However, they are capable of higher working

Self-propelled sprayer

pressures up to 1,000 psi (6,895 kPa). High pressure is used for broadcast or band applications. High pressure is used to force spray through dense foliage or to the tops of tall trees.

High-pressure sprayers range from models that have a single nozzle on a handgun to large units with multiple nozzles that are mounted on a boom. High pressure sprayers are strongly built to withstand extreme pressure. Therefore, they are more expensive than conventional low-pressure models. Many high-pressure units have pressure regulators and bypass mechanisms that allow them to be used as low pressure sprayers.

Air-Boom Sprayers — Air-assistance has recently been applied to field crop sprayers. Air-boom sprayers use a blower unit to carry small spray droplets into the target. Benefits include spraying with lower volumes of carrier, to better coverage and reduced drift. Adding air-assist equipment can add about $15,000 to the cost of a standard sprayer. Therefore, use has mainly been confined to high-value specialty crops and vegetables. A low-cost alternative is to use a system that mixes air and chemical at the nozzle. The conversion is done by adding air-assist nozzles, air lines, and an air compressor or pump and controls.

Foggers and Rope Wick Applicators

Foggers and rope wick applicators are not really sprayers, but they are used to apply liquid herbicides.

Foggers — These sprayers apply pesticides, usually insecticides, as very fine droplets called aerosols. A single aerosol droplet is too small to see, but a concentration of droplets is visible, floating in the air like smoke or a cloud. Consequently, aerosol generators are frequently called foggers.

Aerosol particles can be formed by heat, very fine nozzles, air blasts or spinning disks. Foggers may be electrical or driven by an engine. Most engine-powered foggers discharge aerosol vapor via the engine exhaust system.

Continued on next page

MM16633,00025B0 -19-08DEC10-2/3

Farmers have constructed rope wick applicators to suit their specific needs. Some companies also manufacture the equipment. Rope wick applicators are often used with systemic herbicides. Little or no pesticide is wasted during application, since the applicator wipes tall weeds with herbicide as it is driven through the crop. There is virtually no drift or environmental contamination.

Direct Injection

Direct injection systems offer convenience and safety benefits. The basic design is one that holds undiluted pesticide and water or carrier in separate tanks. Undiluted pesticide is metered into nozzle lines by pumps or air pressure for blending with the carrier before It reaches the nozzle. This eliminates the need to add and mix chemicals in the spray tank, so there is no tank contamination or excess mix at the end of the application. The sprayer is cleaned by simply flushing chemical lines.

The chemical may be held in the original container, in a chemical carrying tank, or in a sealed, returnable container. The returnables virtually eliminate all pesticide handling and disposal of empty containers.

Most sprayers can be equipped with a direct injection system. More than one pesticide can be applied In the same operation, even if they are not compatible. Injection systems can be controlled by electronic monitors, so chemical rates are automatically adjusted to meet specific situations. Injection pumps are also used to apply pesticides through irrigation lines.

MM16633,00025B0 -19-08DEC10-3/3

Application of Dry Chemicals

Types and Sizes of Equipment

Wide variations in dry-chemical-application equipment are due to the following factors:

- Rate of application
- Characteristics of the chemicals used
- Placement or distribution patterns
- Crop characteristics
- Time of application — before, during, or after planting

Equipment for applying dry material may be classified in two groups:

- Broadcast applicators
- Band applicators/injectors

These classes may be further divided as follows:

- Machines that apply only chemicals
- Machines that plant and apply chemicals in one operation
- Machines that till and apply chemicals in one operation
- Machines for aerial application

Broadcast Applicators

Four common types of equipment used to broadcast granular materials are:

- Drop-type distributor
- Rotary spreader
- Pneumatic distributors
- Aircraft

Drop-type Distributor — This applicator has a trailer-mounted hopper with a series of holes or "drops" in the bottom. The operator opens or closes the drops, usually with a lever or cable. The application rate depends on travel speed and the rate at which granules flow out of the hopper. Ground wheels drive a feed shaft that runs the length of the hopper. Rotational speed of the shaft affects the material flow rate out of the drops.

Rotary Spreader — This applicator uses one or two spinning disks to distribute granules. They are metered onto disks by a drag chain running through the bottom of the hopper. The application rate depends on the material feed rate, pattern width and travel speed. The distribution pattern depends on rotational speed of the disks and the point at which granules are fed onto disks.

Units may be trailer or truck-mounted, allowing for large hopper capacities of 10 tons (9 metric tons) or more. Large capacity trailer units usually have "walking-beam" tandem axles that flex on uneven terrain.

Pneumatic Distributors — Pneumatic applicators have a centrally located hopper box and air tubes across the unit's width. Air pressure meters granules through tubes. Granules are distributed after striking deflector plates. Uniform metering and distribution allows farmers to broadcast materials that were once banded. Air systems

Broadcast rotary spreader

have become popular for fertilizer application. Other advantages of air distribution systems are central tank filling and easier installation on equipment.

Aircraft — Airplanes and helicopters also apply granules. They cover large acreage quickly, without compacting soil or damaging crops. Aircraft can operate if soil is wet and are used extensively in flooded rice fields.

Band Applicators

Band applicators are nearly always used as attachments to some other piece of equipment, usually a planter or tillage tool. They consist of the following parts:

- Hopper
- Metering device
- Drop tubes
- Openers — chisels, knives, disks, or other placement device

Granular Fertilizer Attachments

Granular fertilizer attachments have one of two types of metering devices:

- Auger type
- Starwheel type

Auger Type — These attachments are available for low, medium and high-rate applications. Specific rates are obtained by changing the rotational speed of the auger. This is the most commonly used metering device. It can be used with large horizontal hoppers that supply more than one row at a time.

Starwheel Type — Hoppers with a starwheel have circular bottoms. Fertilizer is pushed through a gate opening each time one point of the wheel passes the opening. Application rate depends on the rate of wheel rotation and the width of the opening.

Metering Granular Pesticides

Metering of granular pesticides is usually accomplished by:

- Gravity-flow
- Positive-feed

Continued on next page

MM16633,00025B1 -19-08DEC10-1/2

Gravity-Flow — The gravity-flow type is essentially a hopper with a metering hole in the bottom. Discharge is by gravity, but an agitator is provided In the bottom of the hopper to prevent "bridging". Rate of application is controlled to a large extent by ground speed.

Positive-Feed — The positive-feed type has either an auger or fluted-feed metering device which rotates in the bottom of the hopper to deliver granules to the discharge openings. Application rate can be adjusted by exposing more or less of the fluted-feed roller to granules in the hopper, or changing speed of rotation of the auger or fluted-feed roller and size of the discharge opening.

Positive-feed applicators are more complicated and costly than the gravity-flow type. However, they are more accurate and are less sensitive to changes in travel speed.

The pesticide spreader is located at the end of the discharge tube. It usually has a series of inverted V's inside a plastic or sheet-metal housing.

For control of some pests, such as corn rootworm, the banded pesticide should be worked into the upper one or two inches (2.5 to 5 cm) of soil. Various incorporators are used, including disks, chains, or rollers.

Air Seeders

Air seeders also apply granular fertilizer. They have a central tank, a fan and a chisel plow or field cultivator. The fan blows fertilizer and seed through flexible tubes to boots attached to shanks mounted on the tillage tool. Air seeders can deep-band fertilizer. An optional double fan system places seed and fertilizer separately but at the same time. This allows for increased fertilizer rates. Air seeders are used to plant small grains. soybeans and other crops in rows that are too narrow to cultivate.

MM16633,00025B1 -19-08DEC10-2/2

Application of Anhydrous Ammonia

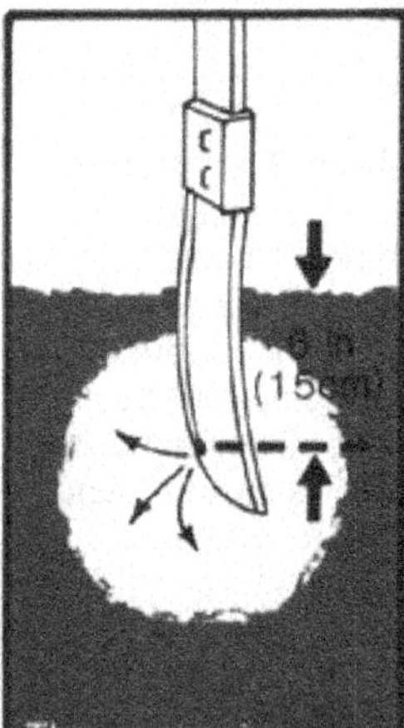

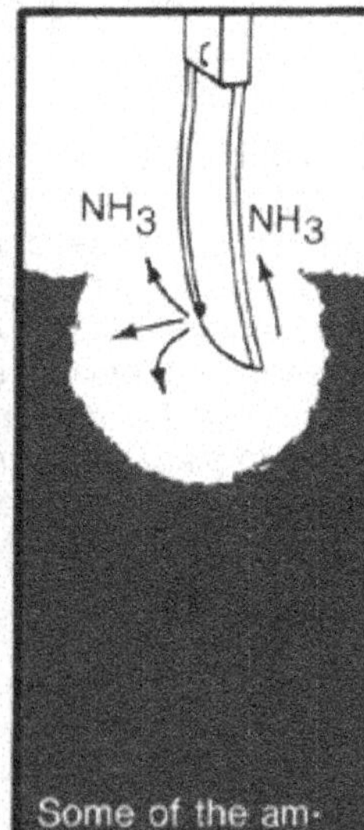

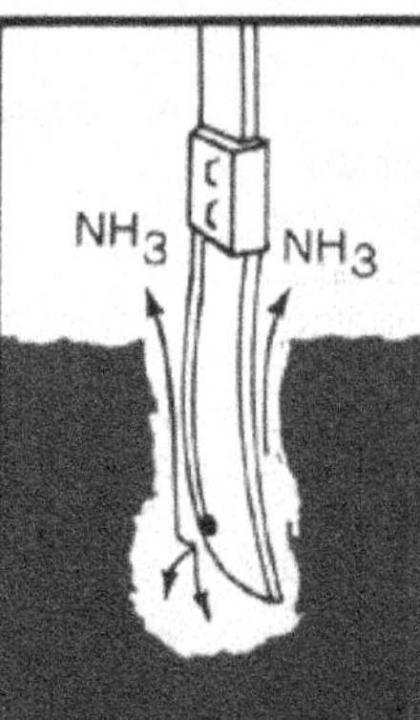

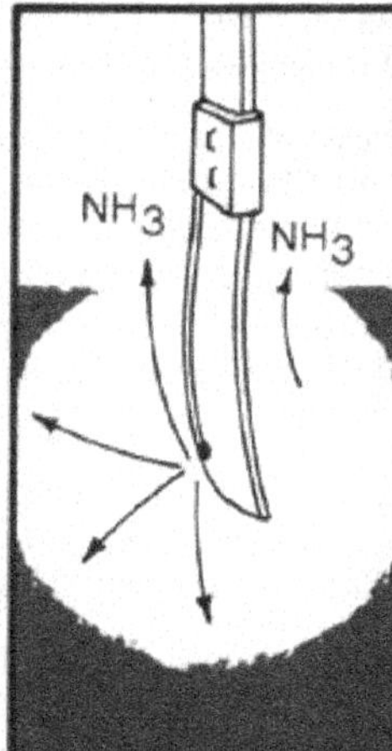

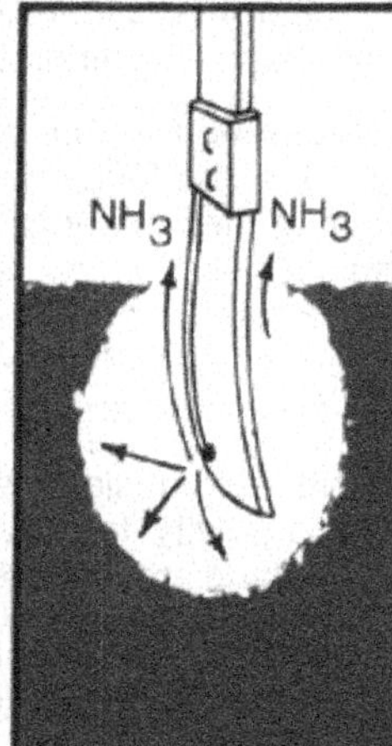

Ammonia placement

Anhydrous ammonia (NH_3) is a popular nitrogen fertilizer source due to its high concentration of nitrogen (82%) and low cost compared to other nitrogen forms. It must be held in a pressurized tank to remain in a liquid form. Because it begins to convert from the highly-concentrated liquid form to the less-dense vapor form as the anhydrous travels through the application unit, it can be challenging to uniformly and accurately meter the product. For the same reason, it cannot be applied on the soil surface and must be injected into the soil to effectively trap the vapors.

Equipment to apply anhydrous ammonia can be a relatively simple design compared to liquid or dry applicators. The tanks used to transport the product are drawn through the field along with a toolbar, similar to a tillage tool, that breaks the soil and injects the fertilizer. But because of the metering challenges previously mentioned, more sophisticated manifolds and applicators have been developed to measure the product and then place it as precisely as possible in the soil before closing the slot.

MM16633,00025D2 -19-09DEC10-1/1

Factors Affecting Water Quality

Any agricultural chemical, particularly nitrogen and phosphorus, may become contaminants of surface and groundwater supplies if not managed properly.

Management practices that minimize the potential for contamination include:

- Apply nitrogen near the time of plant need
- Apply nitrogen and phosphorus at the correct rate
- On sandy soils, apply nitrogen through the irrigation system as the crop is growing
- Take credit for nitrogen from previous legume crop and for application of manure
- Avoid surface application of manure or agricultural chemicals to frozen ground when slopes exceed 3%
- Incorporate manure whenever possible; this also will reduce odors. Incorporating urea fertilizer or soil-applied herbicides can reduce volatilization and improve performance.

MM16633,00025D3 -19-09NOV10-1/1

Introduction

Although many people may know what a combine is, few know the details of what it does or how it works.

The harvester-threshers of yesterday were cumbersome machines that required very experienced operators. And they lacked the capacity, efficiency, and crop flexibility that we have in our present machines.

The modern combine not only performs its basic functions better, but it also makes operation easier, safer, and more comfortable because:

- Convenient controls allow the operator to change speeds and settings from the operator's seat.
- Hydraulic power allows the operator to move heavy loads by simply moving a lever.
- Monitoring systems permit the operator to check shaft speeds, concave clearance, cleaning shoe settings, and even crop moisture content, current yields, and separating losses.
- Operator's cab provides temperature control to keep the operator comfortable in any kind of weather and shields them from machine noise. Automated steering systems have been added to reduce operator fatigue and increase machine efficiency. Video monitors in the cab allow the operator to view items such as straw chopper spread and unloading auger operation.
- Refinements have been made to make the combine as trouble-free as possible. Not only has operation been made easier, but maintenance has been reduced. The development of the combine increased the harvesting of wheat, for example, from fractions of an acre per day with hand tools to 75 or more acres a day with modern combines.

Modern combine

Combine Types and Sizes

Modern combines are available in a wide range of types and sizes. Technology has given us a machine that is very flexible — a combine that can harvest many types of crops under various field and crop conditions.

Self-propelled combines can be described as either level-land combines or hillside combines. Self-propelled combines are available in a variety of models, depending on the crop to be harvested.

Combines may also be classified as conventional, with cylinder threshing and straw walker separation systems, or as rotary (sometimes called axial), in which a rotor or rotors replace the cylinder and straw walkers for threshing and separation of grain and straw.

For more information regarding combine harvesting refer to Combine Harvesting FMO15105NC.

MM16633,0002565 -19-02DEC10-1/1

Operation of Components

Today's combine is a complex machine. Not only are the harvesting and threshing units complicated, but add to that the engine, power train, electrical system, and hydraulic system, and it becomes one of the most complex machines in agriculture. To understand the operation of a combine, look closely at each function of the machine. Once the operation of each of these components is understood, it becomes easier to understand how they relate to each other and the operation of the entire machine.

Cutting Platform Operation

Depending on the crop, a combine may be equipped with either a regular cutting platform, which is used in most crops except corn and rice, or it may have a "draper" platform, which is used in rice. The draper platform is similar to the regular cutting platform except it has a draper, or conveyor belt, between the cutterbar and the auger. The draper aids in picking up or getting more crop into the combine. Cutting platforms vary in width.

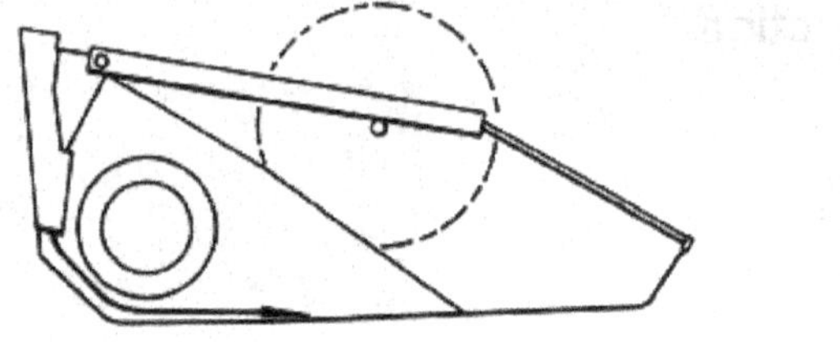

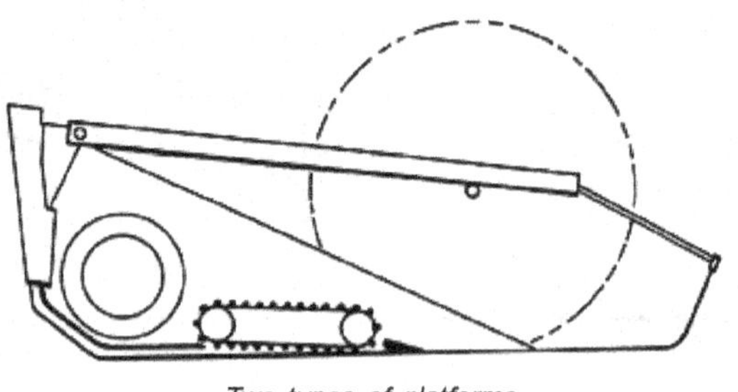

Two types of platforms

Continued on next page

MM16633,00025B2 -19-11NOV10-1/14

102615
PN=78

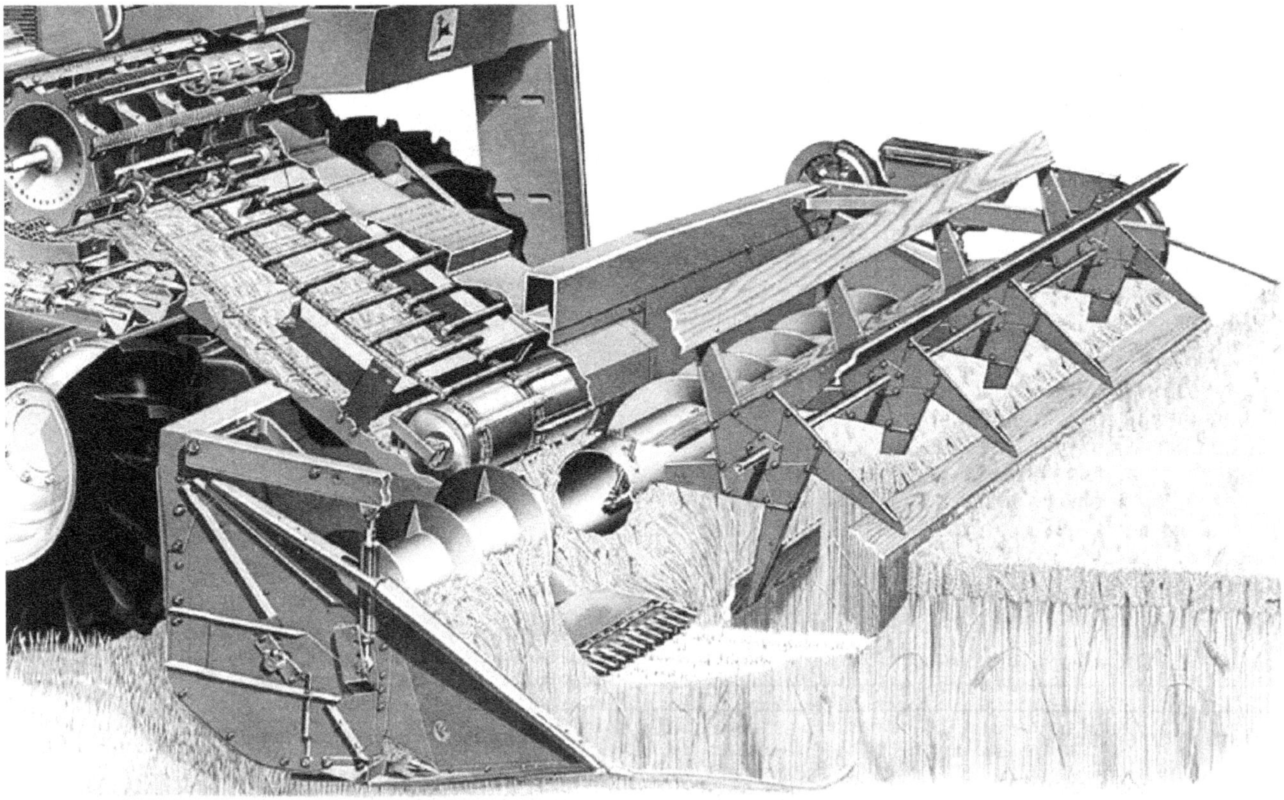

Cutting platform operation

As the combine moves forward in the field, the dividers and end sheets separate a swath from the rest of the crop. The reel parts a section of the crop and pushes it against the cutterbar. As the material is cut by the knife on the cutterbar, the reel continues to push the crop or lift it into the path of the spiralled auger. (On draper platforms, the reel lifts the crop onto the draper, or conveyor, which carries the material to the auger.) The auger moves the material to the center of the platform where the feeder conveyor delivers it to the cylinder for threshing.

The reel, cutterbar, auger, and feeder conveyor must work in proper relationship to cut and feed the crop evenly to the threshing cylinder without losing the kernels or seeds.

MM16633,00025B2 -19-11NOV10-2/14

Reel Operation

Two types of reels are available when combining various crops under different conditions:

- Bat-type or slat-type reel
- Pickup Reel

The bat or slat reel consists of three to eight slats made of wood or steel. The slats rotate against the standing crop to hold it until the crop is cut by the knife on the cutterbar. Then the slat lays the crop back into the path of the auger.

Slat-type reel

Continued on next page

MM16633,00025B2 -19-11NOV10-3/14

The pickup reel has several tines or "fingers" attached to the slats. The fingers pick up crops which have been blown down or have become badly tangled, such as rice or barley. A slat-type reel without fingers cannot pick up crops in these conditions.

The fingers on the pickup reel reach down into the crop and lift it so the cutterbar can get under it. The pickup reel is also used in very ripe crops, such as soybeans, because a slat-type reel would knock the beans out of the pods, causing heavy crop losses. The fingers of the pickup reel tend to gather the ripe crop gently rather than batting it into the auger.

Both types of reels are usually adjustable. The slats on the slat-type reels and the fingers on pickup reels may be adjusted to enter the crop at the proper angle. This adjustment helps prevent shattering of the crop and permits uniform delivery of the crop to the platform auger. This adjustment is very important on a pickup reel that is being used in a down or tangled crop.

Pickup reel

MM16633,00025B2 -19-11NOV10-4/14

Cutterbar Operation

The reel holds the crop against the cutterbar as it is cut. The cutterbar consists of a steel angle bar on the front of the platform. Attached to the bar are stationary guards with slots through which the knife (sickle) moves back and forth, shearing or cutting the crop. The knife is made up of several triangular blades (called knives or knife sections) that are riveted or bolted to a flat steel bar.

Cutterbar

Continued on next page

MM16633,00025B2 -19-11NOV10-5/14

One end of the knife is connected to a reciprocating drive mechanism which causes the knife to move back and forth at several hundred strokes per minute. Some wide headers have two knives, each one half the length of the header, with a timed knife drive at each end of the header. This permits faster knife operation and reduces knife weight per drive unit.

Wear plates and knife hold-down clips control the relationship of the knives to the guards and complete the cutterbar assembly.

To cut properly, the knife must be sharp and must run smoothly in the guards. Every knife section must rest on its guard to make a shear cut. This means that the guards, wear plates, and knife clips must be in good condition and set correctly. The knives must also register properly with the guards as they move back and forth.

Failure to achieve a clean shear cut results in a tearing or chewing action at the cutterbar. This needlessly agitates a ripe crop, shattering kernels or seeds, and possibly results in large crop losses. It also causes increased cutterbar plugging.

Reciprocating knife drive mechanism

MM16633,00025B2 -19-11NOV10-6/14

A cross section of the cutterbar is shown. Notice how closely the milled slot conforms to the section. The section is about 1/8 inch (3.2 mm) thick and the slot is about 3/16 inch (4.8 mm) wide. This helps prevent material from wedging between the cutting edges.

Occasionally, when the cutterbar fails to operate properly, it must be adjusted.

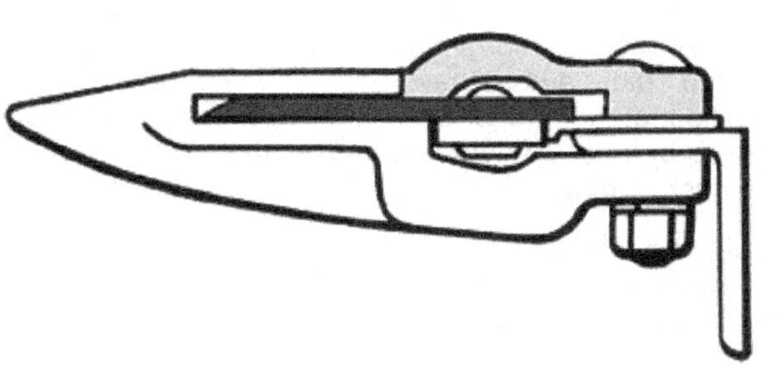

Cutterbar in cross-section view

Continued on next page

MM16633,00025B2 -19-11NOV10-7/14

PN=81

Flexible Floating Cutterbar Operation

The same basic header and cutting parts are used on headers with either rigid or flexible cutterbars. However, instead of being rigidly attached to the front edge of the header, skid shoes and a special linkage permit the flexible cutterbar to closely follow ground contours. This permits cutting much closer to the surface across the full width of the cutterbar. More seeds are saved in crops, such as soybeans, that usually have pods growing very close to the ground. The bar may be locked up for rigid operation in crops such as sorghum or small grains. These flex cutter bars usually flex about 4 inches. (102 mm).

Draper Operation

Drapers (conveyor belts) are used on special platforms for harvesting rice. Because rice is normally a "down" crop when harvested, gathering, cutting, and feeding are difficult to do with a regular cutting platform. The draper platform uses a pickup reel which actually picks up the rice with steel fingers so that the cutterbar can cut it. After

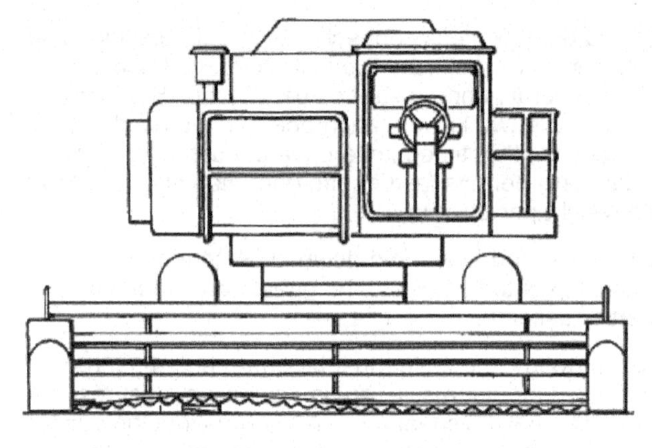

Flexible floating cutterbar

the rice is cut, the reel deposits it onto the drapers, which feed the material to the platform auger.

MM16633,00025B2 -19-11NOV10-8/14

Platform Auger Operation

After the crop is cut by the cutterbar, the reel lays the material back onto the floor of the platform. Here spiral flights of the platform auger sweep the crop toward the center of the platform where the feeder conveyor is located. Most augers have retracting fingers that move the material to the feeder conveyor for feeding the threshing cylinder. (Augers on draper platforms do not have these fingers.)

Smooth, even feeding of the material by the auger is essential for proper delivery to the feeder conveyor. The auger is adjustable to achieve correct feeding in most crops and conditions.

Platform auger

Continued on next page

MM16633,00025B2 -19-11NOV10-9/14

PN=82

Belt Pickup Platform Operation

Pickup platforms are used to gather crops which have been cut and gathered into a windrow by a grain windrower. The pickup platform is essentially the same as a cutting platform, except it does not have a reel nor is a cutterbar required. It is equipped with a conveyor belt to which steel or plastic fingers are attached. These fingers pick up the windrowed crop and deliver it to the platform auger.

A cutting platform may be converted to a pickup platform by installing a belt-pickup attachment.

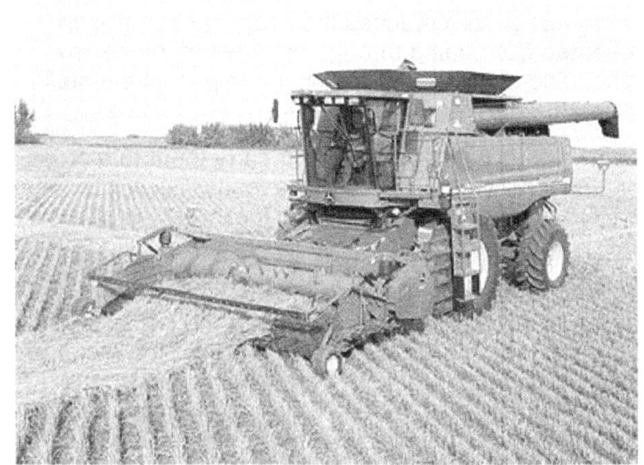

DXP01580 —UN—08JUN10

Belt pickup platform in operation

MM16633,00025B2 -19-11NOV10-10/14

Corn Head Operation

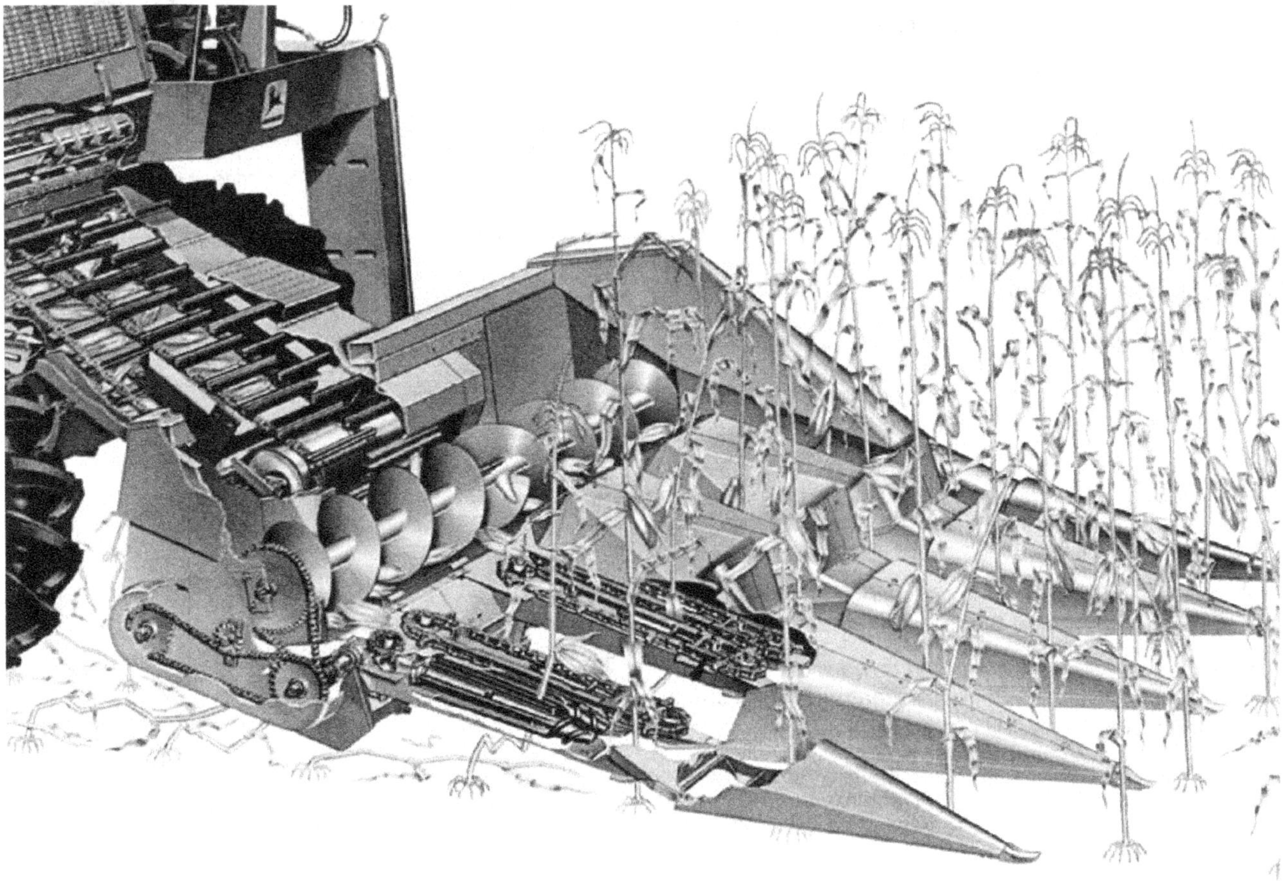

DXP01569 —UN—08JUN10

Corn head in operation

The corn head is essentially a corn picker mounted to the feeder conveyor of the combine. Corn heads vary in size.

As the combine moves through the field, the gatherer points are positioned between the rows of corn. Snapping rolls grab the corn stalks and pull them rapidly down between the rolls.

Continued on next page

MM16633,00025B2 -19-11NOV10-11/14

When an ear of corn reaches the snapping bar, the ear is prevented from going through because of the narrow opening. The snapping rolls continue to pull on the stalk and snap the ear free of the stalk.

Gatherer chains catch the ears and carry them to a cross auger which delivers the ears to the feeder conveyor. The feeder conveyor delivers the ears to the threshing cylinder.

The snapping rolls must operate at a speed in direct relationship to the forward speed of the combine in order to pull the stalks through the rolls before the combine runs over them. If the snapping rolls are operated too fast, the ears may bounce off the corn head and be lost on the ground. High speed may cause shelling at the snapping bars which will also result in losses. It may also cause the entire stalk to be pulled into the combine, resulting in overloading the machine.

If the speed is too slow, the ears will be snapped at the back of the rolls, causing congestion and possible plugging of the head.

The corn head must operate close to the ground to pick up low ears. The gatherer points are adjustable for various crop or field conditions.

Some corn heads are equipped with additional knife blades below the stalk rolls for additional chopping of the corn stalks.

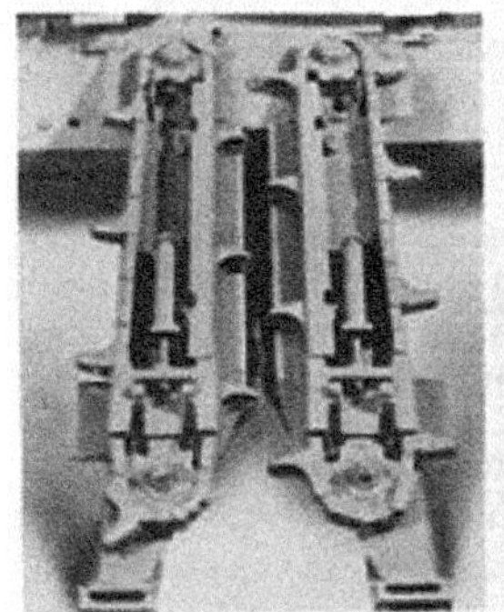

Gathering mechanism

Additional knife blades

MM16633,00025B2 -19-11NOV10-12/14

Row-Crop Head Operation

Row-crop heads are used in crops such as soybeans, grain sorghum, sunflowers, and similar crops to reduce cutting and gathering losses. The gathering points are very low and sloping so that they lift and guide lodged or tangled crops into the gathering area.

Rubber gathering belts on each row unit hold the plants as they are cut off by a knife. The knives are located just below the forward end of each set of gathering belts.

After the plants are cut, the gathering belts convey the crop to the cross auger at the rear of the header.

Each row unit is free to float up and down within a preset range to follow ground contour. This feature helps permit cutting of soybeans that grow close to the ground.

By eliminating a reciprocating knife and rotating reel, seed shattering is reduced. This permits faster operating speed than would be possible with a regular or flexible cutterbar header.

Row-crop head

Continued on next page

MM16633,00025B2 -19-11NOV10-13/14

Feeder Conveyor Operation

The movement of the material from the platform or corn head is done by the feeder conveyor chain or the rake in the feeder conveyor unit. The feeder conveyor takes the material from the platform or corn head and feeds it to the threshing cylinder.

In some combines, a feeder beater is used to help move the material from the platform auger to the feeder conveyor. This beater consists of a round drum equipped with retracting fingers that are similar to the fingers used in the platform auger. In one type of feeder conveyor, the feeder beater feeds the material from the platform directly to the cylinder. Some combines may have a series of paddles that move the material to the cylinder. The most common type of feeder conveyor consists of the chain conveyor or a combination of a feeder beater and chain conveyor.

The chain-type feeder conveyor is allowed to "float" at the lower end to permit smooth feeding of both small and large masses of material. The conveyor chain is adjustable for various crop conditions.

A variable-speed feeder control is available for some combines that permits the operator to adjust feeder speed

Feeder conveyor

to crop conditions and travel speed. Some combines also have a reversible conveyor drive to aid in clearing the feeder conveyor in case it becomes plugged.

MM16633,00025B2 -19-11NOV10-14/14

Combine Operating Controls

Almost all the operating controls and instruments are located at the operator's station within easy reach from the driver's seat. The operator's station is located high on the front of the combine to give the operator a good view of the header. Here the operator can control all of the combine functions.

Most combines have essentially the same controls and instruments. Their names and locations may be different, but they operate the same functions. For example, the control to raise or lower the header may be called either a platform lift control, a table lift control, or a header lift control.

Levers, knobs, or switches may be used to control the various components. Some manufacturers have controls of different colors and shapes to help the operator quickly identify the controls while operating the combine.

Here are some of the color codes one manufacturer uses for the controls:

- Black — Operating adjustments and controls
- Orange — Ground drive and engine speed
- Yellow — Drive engagement

Identifying Controls

Each combine operator's manual has a section on controls and instruments to help the operator identify and operate the controls. Because each manufacturer locates these controls differently, the operator must become familiar with the controls before attempting to operate the combine. Some combines have only a few of these and others may have different ones.

Combine operating controls

Armrest console controls

Continued on next page

MM16633,00025B3 -19-08OCT10-1/3

PN=86

Symbols for Operating Controls

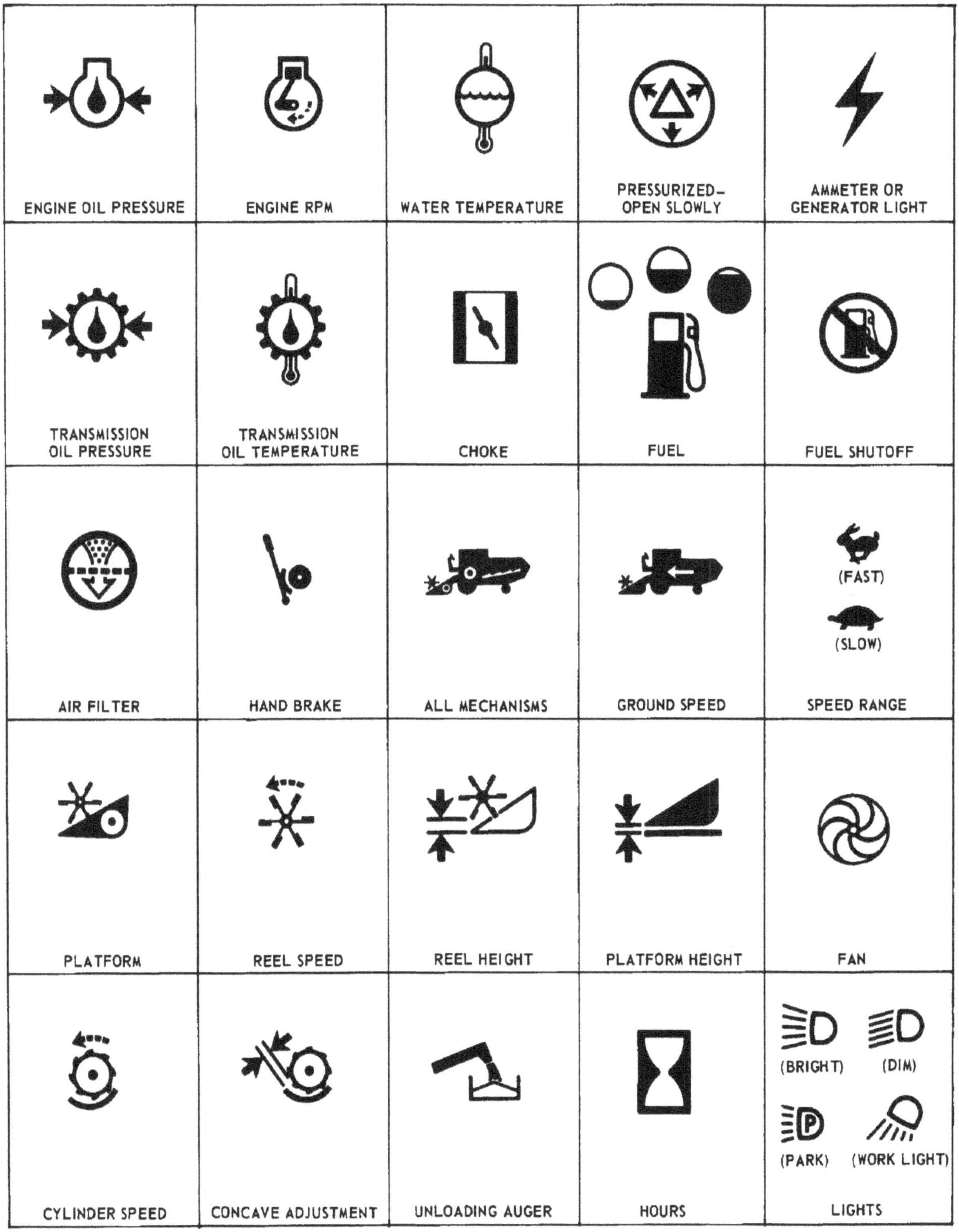

ENGINE OIL PRESSURE	ENGINE RPM	WATER TEMPERATURE	PRESSURIZED— OPEN SLOWLY	AMMETER OR GENERATOR LIGHT
TRANSMISSION OIL PRESSURE	TRANSMISSION OIL TEMPERATURE	CHOKE	FUEL	FUEL SHUTOFF
AIR FILTER	HAND BRAKE	ALL MECHANISMS	GROUND SPEED	(FAST) (SLOW) SPEED RANGE
PLATFORM	REEL SPEED	REEL HEIGHT	PLATFORM HEIGHT	FAN
CYLINDER SPEED	CONCAVE ADJUSTMENT	UNLOADING AUGER	HOURS	(BRIGHT) (DIM) (PARK) (WORK LIGHT) LIGHTS

Universal symbols

DXP01575 —19—05OCT10

Continued on next page

MM16633,00025B3 -19-08OCT10-2/3

Some manufacturers use symbols instead of words to indicate the function of each operating control. Study these universal symbols and learn to recognize them at a glance.

MM16633,00025B3 -19-08OCT10-3/3

Field Operation and Adjustments

Even an experienced operator can learn new tricks

No one can qualify as a good combine operator by simply reading the operator's manual; a competent operator must have experience. Even those who have had experience can always learn new "tricks." Here we will discuss the basic operation and adjustments which can help the operator do a good job of combine harvesting.

To be a good combine operator, you must know:

- Functional design of the combine
- Basic principles of operation
- How to make proper adjustments
- How to identify harvesting losses
- How to maintain efficient operation

Proper Operation and Adjustment Are Important

Combine harvesting can be profitable only if the operator knows how to adjust the combine properly and operate it efficiently with a minimum of losses.

In corn, for example, recent studies have shown that profit losses of 9% or more can result if harvesting efficiency is not ideal. In soybeans, the profit loss can be 13% or more.

What is ideal efficiency? In some studies the average harvesting efficiency for 100-bushel-per-acre (5 t/ha) corn was 92%, while the ideal was 97%. In 30-bushel-per-acre (1.6 t/ha) soybeans, 87% efficiency was average, while ideal was 97%.

How do you get ideal efficiency? It is basically a question of proper operation and adjustment of the combine.

Harvest losses directly result in decreased profits. These losses are caused by poor adjustment and operation of the combine. In corn yielding 100 bushels per acre (5 t/ha), for example, if losses were 4 bushels per acre (200 kg/ha), value lost per acre would be $10.00 at a price of $2.50 per bushel (10 cents per kg). In addition, marketing losses include penalties for low test weight, moisture, damages, and excessive trash or foreign material, all of which can add significantly to the profit losses.

Continued on next page

MM16633,00025B4 -19-09NOV10-1/2

What are proper operation and adjustment?

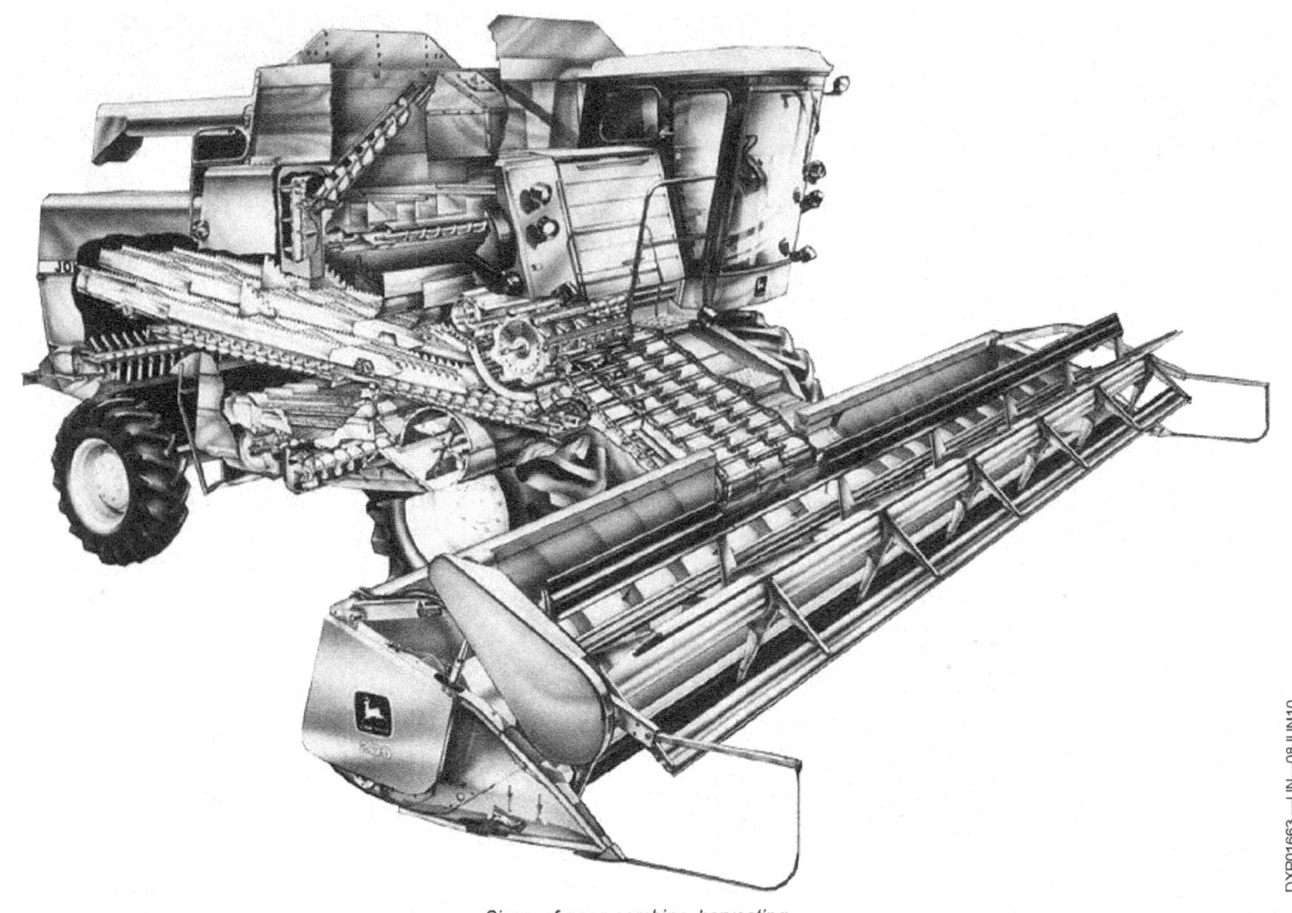

Signs of poor combine harvesting

The operator must recognize the effects of both good and poor operation. Some of the signs of poor combine harvesting are:

- Loose grain losses or whole unthreshed cobs on the ground
- Unthreshed kernels on the straw, cob, or in the pod.
- Straw chewed up excessively
- Grain lost from the straw walkers or shoe
- Excessive tailings in the tailings elevator
- Cracked grain in the grain tank
- Chaff or trash in the grain tank
- Marketing penalties for low-quality grain due to harvesting damage or crop condition

Factors That Affect Combine Harvesting

Several important factors affect the harvesting of crops with a combine. The following factors must be considered in every instance of combining:

- Skill of the operator
- Condition of the crop and field
- Adjustment of the combine
- Proper operating speed of harvesting components
- Ground speed of the combine
- Width of the header

MM16633,00025B4 -19-09NOV10-2/2

Maintenance

Proper maintenance and service adjustments are necessary to ensure efficient, safe operation of the combine.

Costly repairs, premature wear, loss of field time, and accidents can be reduced if the combine is properly maintained and adjusted.

The operator's manual for the machine should be used in reference to specific maintenance intervals, location of service points, and instructions for the performance of maintenance and service adjustments. Always study the operator's manual carefully to determine what maintenance is needed.

There are several practices that a good machine operator always follows. The operator knows that by following these rules, the job of operating and maintaining the combine will be much easier and safer.

1. Always keep the machine clean. Before starting the combine, clean all field trash, mud, and excess grease and oil from the machine. Not only is this a good safety practice, it also helps the combine to run more efficiently, prevents moisture accumulation and rust on metal parts, and cuts down on time lost in the field for repairs.
2. Make sure that nuts, cap screws, shields, and sheet metal parts are tight. A loose shield can vibrate, produce irritating noise, and cause a machine failure if it falls in the way of moving parts. Loose attaching hardware can cause breakdowns that take time the machine should be using for work.
3. Inspect the combine before starting every day. A brief look at all areas of the combine can help you spot potential machine failures and safety hazards.
4. Keep maintenance records. A simple chart showing when lubrication and service adjustments were made can help you make sure that all needed maintenance has been performed.

Keeping maintenance records

5. Don't abuse the machine. Proper lubrication and adjustment is of little help if the operator abuses the machine. A good combine operator follows the operator's manual and doesn't overload the machine, operate it at speeds too fast for field conditions, or operate it under conditions that could cause damage to the machine.

MM16633,00025B5 -19-04OCT10-1/1

Hay and Forage Harvesting

Introduction

Today's hay and forage harvesting machines have greatly improved the efficiency of crop handling and storage. What was once a labor intensive part of farming is now highly automated. This chapter introduces some of the hay and forage harvesting equipment, touches on the how and why of feeding and nutrition, and the selection and growing of hay and forage crops. Good hay and forage management requires an understanding of the machines that handle crops from cutting to storage. It also requires that you preserve as much of the nutritional value in the crops as possible, with the lowest investment of labor and money.

For more information regarding hay and forage harvesting refer to Hay and Forage Harvesting FMO14105NC.

MM16633,00025D4 -19-02DEC10-1/1

Hay

Hay is produced in every state in the union. Millions of tons of hay are harvested each year. The production volume is second only to corn. Hay is a green forage crop harvested for livestock feed and stored at low moisture levels so no special storage structures or preservatives are required. Crops harvested for hay:

- Alfalfa
- Clover
- Sorghum
- Birdsfoot trefoil
- Reed canary grass
- Smooth bromegrass
- Bermuda grass
- Wheat grass
- Canada wild rye
- Timothy
- Russian wild rye
- Native grasses
- Cereal grains such as oats, barley, and wheat

Alfalfa is the most common hay crop.

With proper management, hay returns a good profit. It is produced by commercial hay growers and by stockmen who feed their own hay. High quality hay can have nutrient levels nearly equivalent to many concentrates. Hay is one of the least expensive sources of protein for livestock feed.

Also, hay can be grown on rough terrain unsuitable for other crops. Although outside hay storage is acceptable, inside storage helps preserve quality.

Hay harvesting can be a labor intensive operation. However, with modern equipment, the operation can be mechanized from field to feeding.

Many farmers consider hay a second-class crop because it has not matched the production increases of other crops. However, only a small percentage of farmers have applied the best tillage and fertilization practices to increase production. Even fewer farmers have developed the management skills required to grow and harvest high quality hay. The shortage of reliable farm labor to harvest and handle hay has discouraged some hay growers. Some farmers fear rainy, damp weather that can cause severe quality losses on windrowed hay and square bales left in the field.

Farmers must cure their hay without excessive loss to have an efficient operation.

Curing

Hay is cured in:

- Field
- Barn
- Dehydrator

Field curing — This type of curing is inexpensive, but losses in quality and quantity are greater than barn and dehydrator curing. Timely field operations and good machine performance are needed to minimize losses.

Harvesting Hay

Hay is open to damage from bad weather while it is being field cured.

Barn curing — This type of curing is usually a supplement to field curing. Barn drying is used in humid climates. The hay may be dried with natural air or heated air. At least partial field curing is required for natural air barn drying to prevent spoilage before the hay is completely dry. However, losses, in quality and quantity, are usually less than in field curing due to decreased exposure to weather. But barn curing is not widely practiced because it takes additional labor to stack hay uniformly for even air distribution. And, constructing and operating the drying system costs extra money

Dehydrator curing — This type of curing uses fuel such as natural gas to heat the hay and evaporate moisture. A dehydrator can cure direct-cut hay or wilted hay. This method of curing produces the highest quality hay with minimum quantity losses. However, it only pays to dehydrate the highest quality forages, such as alfalfa.

Harvesting or "packaging" field cured hay is an essential part of most haying operations. You have to harvest a lot of hay to pay for a dehydrator.

Storing

Climate determines the kind of storage used. Types of storage commonly used are:

- Barn
- Temporary cover
- Field

Barn — This type of storage provides the most protection from weather so it preserves the quality of hay more effectively than other storage methods. However, barn storage is also the most expensive method. Small rectangular bales are best for barn storage. These bales stack easily and efficiently in a barn. Larger rectangular bales are stored outside or in pole barns or similar structures. Neither small or large round bales stack effectively, but some hay growers are willing to sacrifice some storage efficiency.

Continued on next page

MM16633,00025F2 -19-09NOV10-1/2

Temporary covers — Temporary covers are sometimes placed over hay to keep it dry. Temporary covers are usually polyethylene sheets, canvas, or nylon tarpaulins. Although expensive, temporary covers protect hay quality.

Field — Field storage is the least expensive method of hay storage. But, because the hay is exposed to sun, wind, and moisture, quality losses may be excessive. However, thoughtful selection of the storage site can help minimize loss of quality. A proper site has good drainage and protection from wind. Large rectangular bales, large round bales and loose-hay stacks are often stored outside. Hay stacks with tightly compacted, sloping surfaces resist penetration by rain and wind.

MM16633,00025F2 -19-09NOV10-2/2

Silage and Haylage

Green forage may also be preserved as silage or haylage. Most silage and haylage is fed to dairy and beef cattle.

Silage is forage converted into succulent livestock feed through fermentation.

Haylage is a low-moisture (40% to 50%) hay silage which is normally stored in an oxygen limiting storage unit. However, with excellent management and proper preparation, haylage may be stored in a good conventional silo. Haylage is chopped fine so it will pack down and exclude air.

Some crops that are harvested for silage:

- Corn
- Grain sorghum
- Forage sorghum
- Sudan grass
- Sorghum-sudan hybrids
- Oats
- Alfalfa
- Alfalfa-grass mixtures

Corn is the most common silage crop.

Quality Feed

Because silage crops are harvested in a high-moisture condition, leaf losses are less than from hay harvested and field-cured. Saving leaves increases forage quality. With high-quality forage, a balanced ration can be provided with less supplemental feeding. Also the effect of weather on quality is minimized because silage can be harvested in damp weather. Properly stored silage can be kept two years or more with little nutrient loss.

The silage fermentation process kills weed seeds and helps reduce weed problems on the farm. In addition,

Corn is the major crop grown for silage

stored silage is less susceptible to fire than barn-stored hay.

Limitations

Even with its many advantages, proper management of the entire silage crop from planting through feeding is needed to ensure quality. Management problems encountered with silage are generally associated with harvest and feeding. A silage harvesting operation must be well planned to avoid hauling bulky, heavy, chopped forage long distances from the field to storage. Also, silage must be fed soon after it is removed from storage or it will spoil.

A silo is needed to store silage. If you stop feeding silage, alternative uses of the silo are limited. Also, due to its high moisture content and perishability, there is little flexibility in marketing excess silage. Long-distance transportation is not feasible. A large investment is required for harvesting equipment.

Continued on next page

MM16633,00025F3 -19-04OCT10-1/2

Harvesting

There are two types of silage harvesting operations:

- Direct-cut
- Wilting

Direct-cut — In direct-cut harvesting, a forage harvester cuts the standing crop, then chops and blows it into a forage wagon or truck for transportation to storage. Harvest losses are minimized because the green crop is fed directly into the forage harvester. Also, harvest labor is reduced with an operation that cuts, chops, and blows forage into a wagon or truck in one operation.

Wilting — Wilting starts with windrowing green forage. After the windrowed crop has wilted to about 60% to 70% moisture, a forage harvester with pickup attachment picks up the windrow and chops and blows the forage into a wagon or truck for transportation to the silo. Wilting the crop before chopping increases labor and field losses.

Forage crop wilting is required to reduce the moisture content so proper fermentation occurs to produce silage or haylage. Some crops have too much moisture. Excess moisture will increase seepage loss during storage and cause silage to be soggy, sour, and unpalatable. A wilting period reduces the moisture content enough to overcome these problems.

To produce haylage, the crop moisture must be reduced from 70% to 85% moisture to between 40% and 50%. Curing requires approximately 4 hours in good drying weather.

Silage Additives

Additives are sometimes added to green, chopped forage to aid the ensiling process or to increase feeding value. Additives are:

- Chemical additives
- Feeds
- Nutrients

Chemical additives — Chemicals are added to change or maintain the pH level for proper fermentation. Chemicals

Wilted-forage harvesting

do not add food value nor reduce the silage moisture content. They are usually organic acids, water soluble salts, or finely-crushed limestone.

Feeds — Certain feed additives aid the ensiling process and add food value to the silage. Typical feed additives are molasses or grain. Molasses provides sugar for the ensiling process but does not reduce the moisture content. Grain reduces silage moisture and assists fermentation.

Nutrients — Nutrient additives increase the feeding value. A proper additive mixture can provide a balanced ration for a specific feeding program. Nutrient additives are not intended to aid the ensiling process, and may actually hinder fermentation under some conditions. Nutrient additives reduce or eliminate the need for supplemental feeding. They include mineral supplements, urea, and anhydrous ammonia.

MM16633,00025F3 -19-04OCT10-2/2

Storing Silage

Green, chopped forage, harvested for silage, is stored in silos classified as:

- Vertical
- Horizontal

Vertical Silos

Primary advantages of vertical silos are:

- Relatively low storage losses
- Easy adaptability for automated feeding
- No need for extensive packing
- They can be located near livestock
- Acids formed during fermentation may cause deterioration of some silo walls

Primary disadvantages are of vertical silos are:

- Need for special equipment to fill and unload
- Difficulties in feeding if unloader malfunctions
- High cost of storage per ton of dry matter

Vertical silos may be classified as:

- Oxygen-limiting
- Conventional

Oxygen-limiting — These units can reduce storage losses and are used to store haylage and silage.

Silage and haylage are stored in vertical silos

They cost more per cubic foot of storage space than conventional silos. They are usually made of metal with an inner liner of fused glass, but some concrete silos may be satisfactorily sealed with an interior coating or lining.

Conventional — These silos are constructed from metal, concrete, or tile.

Continued on next page

MM16633,00025F4 -19-04OCT10-1/2

Horizontal Silos

Primary advantages of horizontal silos are:

- No special loading and unloading equipment
- Easy to build
- Adaptable to self feeding
- Cost less per cubic foot of capacity, compared to vertical silos

Primary disadvantages of horizontal silos are:

- Require extensive packing
- Not suitable for automated feeding
- Well drained locations are not always available
- Unless very good management is applied, horizontal silos often incur more spoilage and leaching losses than vertical silos

Horizontal silos may be classified as:

- Bunker
- Trench
- Stack
- Polyethylene bags

Bunker — This type of silo is placed above the ground surface and usually has concrete or plank walls and a concrete floor.

Trench — A trench silo is dug into the ground, usually into the side of a hill. It may have concrete walls and floor, but frequently the soil is used for both sides and floor.

Bunker silos provide economical storage

Stack — A stack silo is merely a compacted pile of silage placed on the ground with no walls. Stack silos should be used only for emergency storage of excess silage. Stack silage should be fed fast, before it spoils.

Polyethylene bags — Silage can also be unloaded and compacted into large, heavy duty polyethylene bags. Compaction and sealing of the end of the bag provide for storage that is relatively airtight.

MM16633,00025F4 -19-04OCT10-2/2

Forage Crop Planning

Growing, harvesting, storing, selling, and feeding a quality forage crop is difficult. In planning a profitable forage system, a good manager must evaluate many factors and details — always keeping in mind: The primary reason for incorporating a forage system into a farm enterprise is to earn maximum profits.

Some factors that must be considered are:

- Suitability of the farming enterprise for incorporating a forage system (soil type and terrain, crop rotation, scheduling harvest operations, etc.)
- Managerial abilities and desires
- Type and yield of crops suitable for the terrain, soil, and climate
- Selecting a forage crop that produces acceptable quality and quantity
- Feeding or selling the forage crop
- Agronomic practices required to produce forage (tillage, fertilization, etc.)
- Availability of sufficient water
- Type of insect and disease controls required
- Labor requirement and availability
- Selection of an appropriate harvest system

Hay is perishable. Of the many factors that affect hay quality, timely and efficient harvest is most important. The crop must be cut at the right time, cured properly, and handled with care to produce top quality hay.

Time field operations carefully

Leaves contain most of the food value in hay, especially in legumes such as alfalfa and clover. Consequently, every field operation must be planned out to minimize leaf loss.

Most field losses occur during cutting and through the time when the hay is picked up for packaging. Field operations and equipment are very important in limiting field loss.

MM16633,00025F5 -19-04OCT10-1/1

Hay Conditioning

Mower-conditioners and windrowers mechanically condition hay by cracking, crushing or bruising stems as the hay is mowed. Conditioning permits more rapid moisture loss from inside the stems. Stems and leaves then dry at approximately the same rate.

Hay can also be conditioned as it is cut by spraying a solution of chemicals with push-and-spray bars mounted on the mower-conditioner or windrower. These solutions usually contain potassium or sodium carbonates. They are intended to speed evaporation of water while limiting leaf shattering. Success of chemical conditioning has been less consistent than mechanical, so most conditioning is done mechanically.

Conditioning reduces the time needed for field curing, so increases the likelihood of packaging hay before it can be damaged by rain. Studies show that conditioning may reduce drying time by 30% to 50%.

Conditioned hay has better quality and palatability than unconditioned hay. Leaf loss from shattering is substantially reduced because hay is packaged before leaves become brittle. Reduced exposure to bad weather helps the crop retain color, vitamins and nutrients. Coarse stiff stems are crushed or cracked and become more palatable. From one to four percent of the potential crop yield may be lost during conditioning, depending on the conditioner and crop conditions. However, the actual quantity of hay harvested may increase because of reduced field losses from handling and exposure to weather.

Conditioning is most effective on coarse stemmed, leafy hays. Legume hay is particularly well suited to conditioning because the leaves become fragile when dried. However, conditioning fine-stemmed grass hays may be beneficial, especially where weather might damage hay before It can dry naturally

Mower-conditioners and windrowers essentially combine the functions of a mower and conditioner to convert standing crops into properly conditioned windrows. Then, with normal crop and weather conditions, no further hay processing is needed before packaging.

Mower-conditioners do not reduce the need for timely harvest to produce high quality hay. However, they ease the scheduling problems required for three separate operations of mowing, conditioning and raking. Also, mower-conditioners and windrowers help save money and labor. Maintenance and storage costs are confined to one machine. For large acreages, self-propelled units can eliminate the need for a tractor during mowing and conditioning.

A major difference between windrowers and mower-conditioners is the conveying of hay by windrowers; hay is moved laterally on the platform by an auger or draper before entering the conditioning rolls or dropping on the ground. However, some equipment manufacturers and customers may use the words "mower-conditioner" and "windrower" interchangeably. Windrowers are used primarily for cutting grain are also known as "swathers" in some areas.

MM16633,00025F6 -19-04OCT10-1/1

Mower-Conditioners

Mower-conditioners are used on farms that do not require a large self-propelled windrower. Mower-conditioners are less expensive than self-propelled windrowers and are more maneuverable, a desirable feature in small, irregularly shaped hay fields.

Mower-conditioners are designed to provide efficient conditioning by using full-width conditioning rolls. Most models use inter-meshing urethane or rubber rolls that crimp and crush the hay. Either crushing or crimping rolls are available for some machines. Impeller conditioners are available for other machines.

Cutting with mower-conditioners may be accomplished by sicklebar or by rotary disks.

Mower-conditioners are PTO or hydraulic driven. Most PTO mower-conditioners have cutting widths of 7 to 16 ft (2 to 5 m).

Hydraulic-driven mower-conditioners have a hitch over the platform. The hydraulic drive allows wide flexibility for turning and maneuvering because it is not restricted by a

Pulled mower-conditioner

mechanical PTO drive train. Consequently, the hitch may be adjusted hydraulically into a variety of positions for field operation and transport. Most of these units have 12 to 16 ft (3.6 to 4.9 m) wide cutting platforms.

MM16633,00025F7 -19-09DEC10-1/1

102615
PN=99

Windrower

Windrowers are used on nearly all commercial hay farms and are widely accepted by farmers who grow hay for their own use. Windrowers have more capacity than mower-conditioners so they are preferred by growers in areas where labor is in short supply. Windrowers are also available without conditioners for harvesting grain crops.

Windrowers may be classified by the type of conveying device on the platform:

• Auger Platform
• Draper Platform

Auger Platform — The auger platform handles all hay, but is particularly effective on crops over 5 feet tall (1.5 m). The auger carries hay from the cutter bar to the center of the platform, where it is fed into the conditioner or dropped on the ground.

Draper Platform — The draper platform cannot efficiently handle tall crops because it is not aggressive enough to bend the long stems. Draper platforms are used on hay-grain farms where one machine handles both crops.

Draper platforms have conveyor belts which carry the crop to the center of the platform. Ribs or slats on the conveyor belts help keep the crop moving.

Self-Propelled Windrower

Self-propelled windrowers provide the highest capacity for cutting, conditioning, and windrowing hay. The mounted

Self-propelled windrower

engine provides power for cutting and conditioning the crop and for ground propulsion. The power train to the drive wheels may be mechanical or hydrostatic. Platforms are usually 10 to 16 feet (3 to 4.8 m) wide for windrowers used in hay. Platforms for some grain windrowers are more than 20 feet (6.1 m) wide.

MM16633,00025F8 -19-05OCT10-1/1

Rakes

Rakes were developed to gather newly-cut hay into small piles or windrows to make hay collection easier. Rakes lift mowed hay from the swath and place it in a loose, fluffy windrow with the green leaves inside, protected from the sun's rays. The leaves retain their fresh, green color, and the stems cure. Rakes are also used to windrow straw and crop residue for harvest or burning.

Mower-conditioners and windrowers that cut, condition, and windrow in one field operation have not eliminated the need for rakes. Rakes combine windrows in light hay crops. Windrows are also combined for high-capacity balers, round balers, and any bale wagons that operate best with heavy windrows. Rakes are sometimes used to turn windrows for even exposure to the sun. Turning rain-soaked windrows for uniform drying is also a common practice.

Side-delivery rakes are classified as:

- Parallel-bar rake
- Wheeled rake

Parallel-Bar Rakes

Parallel-bar rakes require a power source to drive the rake reel and bars. They may be ground-, PTO-, or hydraullcally-driven. Parallel-bar rakes may be further classified by the way they are attached to the tractor:

- Trailed
- Rear-mounted

Trailed Rakes — Some trailed, parallel-bar rakes have a gauge wheel in front. This gauge wheel supports part of the rake's weight and provides immediate response by the rake reel to any terrain irregularities.

Other trailed rakes attach directly to the tractor draw bar without a gauge wheel. Without the front gauge wheel, the tractor draw bar carries part of the rake's weight and response to terrain irregularities is not as fast.

Ground driven, trailed, parallel-bar rake

Trailed rakes are usually ground-driven. But hydraulically driven models are available. Ground-driven rakes are easy to hookup, and have a direct relationship between reel speed and ground speed.

Rear-Mounted Rakes — Rear-mounted, parallel-bar rakes are connected to a tractor's 3-point hitch. All of the rake weight is carried by the 3-point hitch during transport, and most of the weight is on the tractor during operation. Caster wheels on the rear help gauge the terrain. These maneuverable rakes are ideal for raking small, irregular-shaped fields. Rear-mounted rakes can be transported easily by raising the 3-point hitch. Rear-mounted, parallel-bar rakes are driven by the PTO.

Continued on next page

MM16633,00025D6 -19-08DEC10-1/2

Wheel Rakes

Wheel rakes are simpler than parallel-bar rakes because no chains, belts, or gears are needed to drive the wheels. The raking wheels are turned by the rake teeth on the ground. Rake teeth break because they are always on the ground, and they may rake rocks and brush into the windrow. Wheel rakes tend to make tight, rope-like windrows that dry relatively slowly. But individual wheel flotation provides clean raking on rough terrain. Also, wheel-rake teeth travel slower than parallel-bar rake teeth so they are more gentle on hay.

Wheel rakes are trailed or rear-mounted. Trailed rakes offer more flexibility for adding extensions and are available in larger sizes, but mounted rakes are more maneuverable.

Wheel rake

MM16633,00025D6 -19-08DEC10-2/2

Square Balers

Baling is essentially a "packaging" operation. The materials that can be packaged with a baler range from high quality hay to crop residues. The popularity of baling is greater than any other hay packaging method. Such wide acceptance has come because of bale size, shape and density. Small bales can be handled for stacking and feeding and are dense enough for efficient inside storage. Larger bales can be mechanically handled with lower labor costs. Loose hay requires twice as much storage space as baled hay. Bale density of 10 to 16 pounds per cubic foot (160 to 256 kg per cubic m) makes long distance transportation economically feasible. This is important to both commercial and larger-scale hay producers.

Baling formerly was primarily a custom operation. Now, most hay farmers own balers. Small, lower-priced PTO-powered balers and recently developed larger balers match the needs of almost all hay producers. Owning a baler lets a farmer time his own haying operations, which is essential for high quality hay. Custom baling is available for the few hay growers with small acreages who cannot economically justify ownership of a baler.

Balers are classified by:

- Power source
- Size of bale produced
- Twin and tie
- Wire tie

Large square baler

All hay balers produced today are field balers with automatic tying mechanisms. Tractor-drawn balers are powered by the tractor PTO or an auxiliary engine mounted on the baler. Some high capacity balers are self-propelled.

Continued on next page

MM16633,00025D7 -19-05OCT10-1/2

PTO-Driven Balers

A PTO-driven baler requires a tractor for both baler operation and forward propulsion. A PTO-powered baler is the least expensive. The tractor must be large enough to maintain a constant engine speed during baling and have enough forward speeds to match crop conditions Almost all balers now in use are PTO driven.

Self-Propelled Balers

Self-propelled balers are used on a few large hay farms. Self-propelled balers are more expensive than other pulled balers. but they do not require a tractor. Both the baler pickup in front and the bale chute behind can be seen from the seat.

PTO driven baler

MM16633,00025D7 -19-05OCT10-2/2

Round Balers

The round baler packages cured windrows into firm, round bales. After the bale is formed, it may be wrapped with twine or surface wrap for stability. Compact, weather-resistant bales are then stored outside or in sheds until they are fed or ground into rations.

Many features are available on large round balers. Some balers are driven mechanically by the tractor PTO shaft. Others are hydraulically driven. The twine-wrapping mechanism is controlled manually on some models, hydraulically on others. Some models have optional surface wrapping. However, the most distinctive difference between models is the baling chamber.

Earlier models of large round balers formed bales by rolling the growing bale on the ground, then lifting the forming mechanism over the bale when it was finished. Current models of large round balers carry the bale in a bale chamber as it is formed, but differ in the method of bale fanning.

Variable-Chamber Balers

With variable-chamber balers, all hay entering the chamber is compressed continuously for relatively uniform density throughout the bale. Bale size can be varied by the operator.

The most common type of variable-chamber round baler uses forming belts that apply pressure on the incoming hay to make the bale as it rotates in the bale chamber. Pressure can be changed to vary bale density.

Variable-chamber round baler

In addition to controlling bale formation by belts, other balers use rollers, particularly during the initial formation of the bale core.

Fixed-Chamber Balers

With fixed-chamber balers, hay at the center of the bale is compressed less than hay on the outside of the bale, Only one size of bale is possible.

Hay enters the chamber until enough mass accumulates to start bale rotation. As additional material enters the chamber, it is packed or formed on the outside of the developing bale until desired density is achieved.

MM16633,00025D8 -19-11NOV10-1/1

Introduction

Precision farming is a term that generated a tremendous amount of interest in the last decade of the twentieth century. In fact, the very definition of precision farming: managing each crop production input — fertilizer, limestone, herbicide, insecticide, seed, etc. — on a site-specific basis to reduce waste, increase profits, and improve the quality of the environment, allows farmers to become even better stewards of their farmland.

This concept of site-specific management-treating small areas of a field as separate management units is not new and actually follows the practices of the world's first farmers who tended individual seeds and plants.

Obviously, with today's large-scale agriculture, the individual treatment of each plant is impossible without some remarkable assisting technologies. It was the invention and availability of new technologies that have made the concept of precision farming feasible.

Continued on next page

OUO1023,00040C3 -19-23OCT15-1/3

A precision farming system

A—Variable Rate Application
B—Crop Scouting
C—Variable Rate Herbicide Application
D—Harvesting with a Yield Monitor
E—Data Analysis
F—Soil Sampling

Today, technology has reached a level that allows a farmer to measure, analyze, and deal with in-field variability that was known to exist previously but wasn't manageable. The ability to handle variations in productivity within a field and maximize yields has always been a desire of the farmer, especially the farmer with limited land resources. Microprocessors, sensors, and other electronic technologies are new tools available to help all farmers reach this goal. The description of this technology and its implementation is the focus of this chapter.

Continued on next page

OUO1023,00040C3 -19-23OCT15-2/3

PN=105

Many farmers will adopt some aspect of precision farming without developing an entire system. However, for those that seek the ultimate rewards of precision farming, the goal is to apply all crop production inputs in site-specific manner. Those farmers identify variability within each field and deal with that variability using variable-rate application (VRA) technologies, variability manifests itself in two basic ways — spatial and temporal. Spatial variability is the variation in crop, soil, and environmental characteristics over distance and depth. Temporal variability is the variation in crop, soil, and environmental characteristics over time. Variability is found in all farm fields. Variability can be seen in soil fertility, moisture content, soil texture, topography, plant vigor, and pest populations.

Some soil-related characteristics are very stable, changing very little over time, like texture and soil organic matter (SOM) content, for example. Other characteristics, such as nitrate (NO_3^-) levels and moisture content, can fluctuate daily. Of course, crop conditions can change in a matter of hours. Precision farming involves collecting soil and crop samples to gain information about how conditions vary in the field. Variability over time affects a number of precision farming decisions including: what variable to sample, how to sample, how often to sample, and how to deal with measured in-field variability.

Sampling methods differ in terms of the expense in collecting samples and in sample analysis. Sampling frequency requirements can affect the way in which the farmer manages money, labor, and time. Some crop production inputs can be varied based on maps generated from sampling data collected months, even years, prior to application. Limestone applied to address soil pH variability is one example of such an input. However, other inputs are based on soil characteristics that change so rapidly that traditional sampling techniques might not be appropriate for prescribing inputs. If a characteristic changes rapidly, it might make sense for the farmer to use application equipment that can sense the variations and respond by varying input rates almost instantaneously, in "real time," without relying on any prior sampling.

Current whole-field management approaches ignore variability in soil-related characteristics and seek to make applications of crop production inputs in a uniform manner. In fact, farmers viewed application controllers that allowed them to maintain constant application rates across the field as "state of the art" not too long ago. Constant-rate applications were often based on properties measured in composite soil samples that were collected to represent the average characteristics of a whole field. With such an approach, there was obviously the likelihood of overapplication and underapplication of inputs in a single field.

So, now that technologies exist to help farmers do a better job of managing inputs, what are the forces driving their adoption? Economics is among the most important factors effecting (and affecting) the transition from whole-field to site-specific crop management. Precision farming has the potential to affect input costs, crop production revenue,

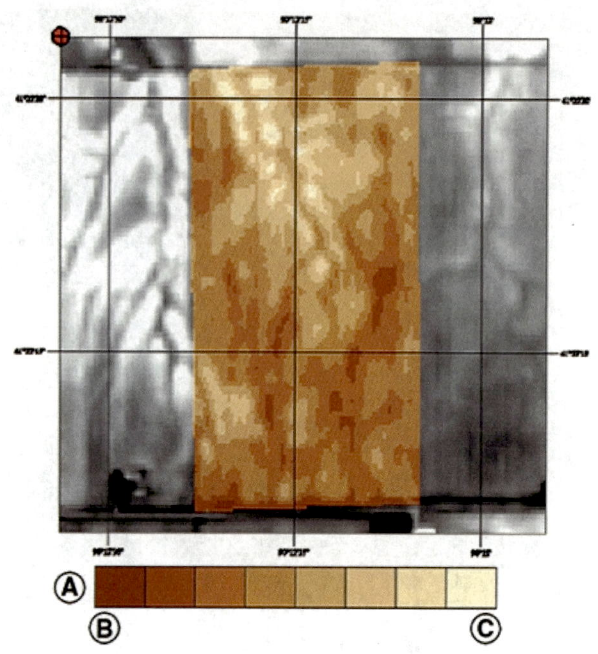

A map showing variations in soil color within a field

A—Soil Color Graph   C—Lighter
B—Darker

and the environmental effects of fertilizers and pesticides. There is the potential for:

- Greater yields with the same level of inputs, simply redistributed
- The same yields with reduced inputs
- Improved crop quality which can boost revenues

The farmer must decide on an appropriate strategy for his or her own operation.

First, a farmer must determine just how much variability exists in his or her fields. The next, and most important, step is to find cause-effect relationships between measured variables and measured output, in terms of both crop yield and quality. This means knowing what each area in a field is capable of producing and what keeps any area from producing to its full potential. Information must be collected to determine yield limiting factors on a site-specific basis throughout each field. If a farmer can find out what is causing a yield or quality difference, can his or her tools and resources fix the problem? Finally, does it make economic sense to use those tools and resources to fix the problem?

The crop production inputs that farmers currently apply in a spatially-variable or variable-rate manner include fertilizer, pesticides, and seed. As you will see, each of these inputs is of great importance to the farmer.

For more information regarding precision farming refer to The Precision-Farming Guide for Agriculturists FP403NC.

OUO1023,00040C3 -19-23OCT15-3/3

Fertilizer, Pesticides, and Seed

Each year the world's farmers apply over 100 million tons of nitrogen (N), phosphorous (P), and potassium (K) fertilizers. In the United States, 97% of all acres planted to corn receive an application of nitrogen fertilizer. For a typical Midwestern corn grower, fertilizer accounts for over one fourth of total cash production expenses. Therefore, the ability to better manage such a high-cost input can have a significant impact on the profitability of a crop production operation.

Farmers who implement precision practices might see higher crop quality resulting from better matching of inputs to crop needs. It is well known that nutrient deficiencies can reduce crop growth and lower crop quality. Over-application of fertilizers can also reduce yields and crop quality. For instance, over application of nitrogen can increase vegetative growth but reduce grain production in crops such as wheat and rye. Too much applied or residual nitrogen can reduce sucrose content in sugar beets.

In addition, the farmers consider the potential for environmental impacts. Nitrates that leach through the soil often end up in groundwater — the same resource that provides drinking water for over 80% of rural Americans. Phosphorous makes its way into lakes and streams as a soil-bound contaminant, carried by eroded soil particles. Excessive levels of N and P in bodies of water can harm many forms of aquatic life. If N and P applications can be matched to crop and soil needs, agriculture's contribution to water pollution should decrease as a result.

It is desirable to place fertilizers in the correct place at the correct application rate, as determined through agronomic analysis. To date, most VRA research, development, and commercialization has taken place in the area of fertilizer application. At times, it is desirable to apply combinations of fertilizer. It is possible to change combinations of fertilizer "on-the-go" as an applicator travels across the field.

Each year in the United States, farmers spend over $11 billion on the purchase and application of agricultural chemicals — herbicides, insecticides, and fungicides. Ninety-eight percent of all acres planted to corn and soybeans receive herbicide applications. Improper application of pesticides can have negative effects during the crop growing season and well beyond. If application rates are too low, pest control is poor. If application rates are too high, pesticides can be toxic to the crop, can carry over to future growing seasons, and can end up in groundwater. Variable-rate application of pesticides holds the potential to save significant sums of money and reduce the potential for crop and environmental damage. There have been reports of variable-rate technologies leading to application reductions of 50% and more. One can easily imagine that if pesticide is sprayed only on weed targets in a field instead of being broadcast on all plants and between rows of plants, significant amount of pesticide can be saved.

The great increases in crop production in the United States during the 20th century (average corn yields up 2% per acre per year since 1948) can be attributed, at least in part, to the development and widespread use of high-yielding hybrid varieties of crops. In 1900, one farm worker in the U.S. produced enough food and fiber for eight people. Along with the use of chemical fertilizers, pesticides, and improved field machinery, ever-improving crop varieties now enable a single American farmer to provide food and fiber for well over 100 people around the world. Some of these crops are sensitive to plant spacing and plant population. The technology now exists to accurately place seeds and to vary seeding rates to match the productivity or yield potential of soils across the country.

MM16633,00025F9 -19-11NOV10-1/1

102615
PN=107

Satellite-Based Positioning Systems

A positioning system, as the name suggests, is a general method of identifying and recording the location of a stationary object or moving vehicle or person. Such a system may be used to chart a vehicle's progress along the earth's surface, in the air, or in space. The systems are of great benefit in modern agriculture. In fact, it may be considered the foundation of precision farming by helping farmers accurately identify and record the location of a machine in the field. Yes, it's true that even without precision farming, the farmer must be able to identify his or her location during planting, cultivating, harvesting, and so on. But, the ability to electronically record the vehicle's precise location during field operations has helped implement the concept of precision farming. For example, recording a combine's position every few seconds during harvesting along with data from sensors provides essential data for making crop yield maps.

Systems to estimate the acreage covered by tractors and combines have been available for a number of years. However, the chances for error and the accumulation of errors over time make such systems unacceptable for precision farming operations. Today, satellite based positioning systems are the only method used to help navigate and record positions of agricultural vehicles during field operations.

This section will cover the satellite-based positioning system known as the Global Positioning System (GPS).

GPS is the most common and the most significant positioning system for precision farming and many other civilian applications. Since position accuracy is the main issue with any system, we discuss the accuracy users demand from a positioning system, especially for various farming activities. Users employ certain methods to make positioning systems as accurate as possible. We describe common ways of doing so, including a method to reduce position errors using a process known as differential correction. This procedure results in what is called a Differential Global Positioning System (DGPS). Finally, we cover the types of GPS hardware developed for agricultural use, describing their specifications and applications in precision farming.

MM16633,000257A -19-04OCT10-1/1

Global Positioning Systems

GPS is a satellite-based navigation and radio-positioning system created and operated by the United States Department of Defense (DoD). The system was originally designed to serve as a worldwide navigational aid for the U.S. military, but now GPS serves industrial, commercial, and civilian interests as well. This service is available free of charge, 24 hours a day in all weather conditions. To better describe GPS, we can divide the entire system into three segments:

- Space segment
- Control segment
- User segment

Space Segment

The United States Space satellite constellation is called NAVSTAR (Navigation by Satellite Timing and Ranging). There are over twenty-four satellites that make up the complete constellation. The arrangement of satellites are arranged to guarantee that at least four satellites will be "in view" of your GPS receiver anywhere in the world, 24 hours a day.

Of course we cannot actually see the satellites in orbit, but a GPS receiver must be able to pick up the satellites' signals sent to earth. The only satellites from which signals can be received are those that are "above the horizon" since the signals travel by line-of-sight and cannot be seen around the curvature of the earth.

Each satellite is equipped with radio transmitters and receivers for sending and receiving radio waves.

The satellites are also equipped with atomic clocks which keep time based on natural periodic vibrations within atoms. These incredibly precise clocks are the critical components that make it possible to use the satellites for mapping and navigation.

GPS constellation of 24 satellites (minimum) commonly described as the Space Segment

Control Segment

The GPS satellites are tracked and monitored by several facilities strategically located around the world. This network of monitoring stations is usually referred to as the control segment of GPS. A Master Control Monitor Station is located at Schriever Air Force Base (formerly Falcon AFB) in Colorado Springs, Colorado. Monitoring stations measure the radio wave signals that are continuously transmitted by the satellites and relay information to the Master Control Monitor Station. The Master Control Monitor Station uses this information to compute clock errors and the exact orbits of the satellites. Corrective information is then relayed back to each satellite to update their navigation signals.

Continued on next page MM16633,00025FA -19-09NOV10-1/2

User Segment

The user segment includes the GPS receivers located on vehicles or carried by hand which are used by civilians and military personnel who receive the GPS signals. Civilian GPS receivers do not require a license to operate because they to do not send out or transmit radio signals. They only receive signals. Also, there is no direct charge or fee for using the basic GPS satellite signals. However, as we will see later, GPS receivers vary in design which affects their cost and accuracy.

User segment of GPS includes the receivers located on vehicles or carried by hand

MM16633,00025FA -19-09NOV10-2/2

Yield Monitoring

In an effort to learn as much as they can about what affects crop production and profits, many farmers monitor crop yields in their fields. Traditionally, farmers measured crop yields for whole fields or for large sections of fields.

The mechanization of agriculture and the trend toward larger equipment meant farmers could cultivate larger areas. Unfortunately, many continued to treat their larger fields as single management units. Yet, for years farmers and researchers have documented variability in soil properties, environmental conditions, and crop yields. Soil and environmental variability inevitably affect crop production. Recent technological advances and data processing improvements allow farmers to address this issue.

A growing number of crop producers today practice site-specific crop management (SSCM). Crop yield monitoring is often the first step in developing their SSCM or precision farming programs. Precise crop yield data can be combined with soil and environmental data of many kinds to begin the process of developing a precision crop management system.

Methods for Measuring Crop Yield

There are a number of ways to measure crop yields. Grain yields are most often expressed in terms of bushels per acre. This means there must be some way to associate measured quantities of grain with measured harvested areas in the field. Of course, grain moisture content has a large impact on measured yields. Different moisture contents will cause grain samples of equal volumes to weigh different amounts. Grain yields are, therefore, stated

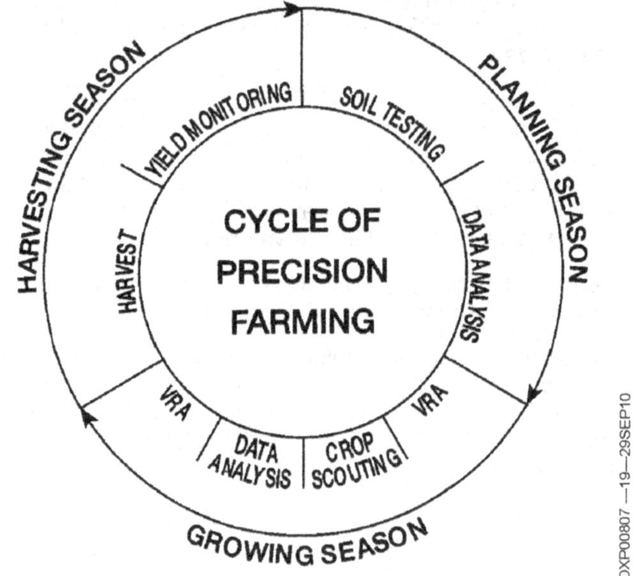

The cycle of processes included in a precision crop management system

in terms of volume per unit area at a specific moisture content. It should be noted that even though a bushel is a measure of volume, crop yields in the United States are measured by weight. A bushel weight is either assumed (56 lb/bu for shelled corn, for example) or measured.

Three major yield measuring approaches are listed below. The first method is the oldest, and is still in use. The second method is considered the forerunner of modern, site-specific yield monitoring.

Continued on next page   MM16633,000257C -19-08DEC10-1/11

Collect-and-Weigh

Used for many years, the collect-and-weigh method determines yields for whole farms, for individual fields, and for harvested strips within fields. Scales at a grain receiving facility or scale-equipped wagons in the field weigh the crop harvested from large areas. Many farmers weigh and record the weight of each wagon or truckload of grain harvested from a field. Crop moisture content is typically measured by sampling each weighed load.

It is possible to make yield maps based on collect-and-weigh data, but there must be some way to measure the area from which each load was harvested. Most yield records based on the collect-and-weigh method are field-scale records. This means that a single average

Grain yield measurement using the collect-and-weigh method

yield value is assigned to an entire field. The average yield value is determined from the total amount of weighed grain harvested from the total planted area of a field.

MM16633,000257C -19-08DEC10-2/11

Batch-Type Yield Monitors

A batch-type yield monitor weighs grain in a wagon in which the grain is loaded or as the grain tank of the combine is unloaded. The weight measurement may be displayed to the combine operator on a monitor in the combine cab.

A typical combine grain tank holds a relatively large volume, or batch, of grain harvested from a relatively large area. For instance, the corn filling a 200-bushel grain tank will generally have been harvested from an area of between one and two acres. As with the collect-and-weigh method, the area harvested must be measured or estimated in order to calculate yield. Though yield values determined from a batch-type monitor are not truly site-specific, grain weight estimates are available to the farmer quickly, and without the need for a separate weighing operation. As yield monitoring technologies have advanced, batch-type systems have given way almost entirely to instantaneous yield monitors.

An early in-cab batch-type yield monitor display

Continued on next page   MM16633,000257C -19-08DEC10-3/11

Instantaneous Yield Monitors

Instantaneous yield monitors measure and record yields on-the-go. A number of different methods are used to measure crop yields on-the-go. On-the-go measurement simply means that the process is continuous as the grain is being harvested. Data points are continuously collected as the combine operates. Some systems record each data point separately. Others collect a number of points that are then processed to provide load summary data. Some systems measure crop volume directly while others weigh the crop. All systems have the capability to measure the area harvested for each recorded weight or volume. When combined with positioning systems such as DGPS, instantaneous yield monitors provide the central data for generating site-specific **yield maps**. Yields are associated with specific locations within a field automatically. Most site-specific yield monitors also measure grain moisture content on-the-go. The collect-and-weigh method is typically used in conjunction with instantaneous yield monitoring for the purpose of calibrating the on-the-go monitors to ensure accuracy.

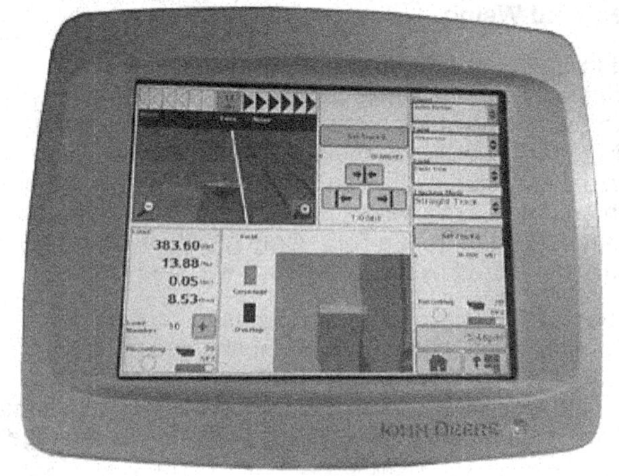

Instantaneous yield monitor displaying crop yield and moisture content

MM16633,000257C -19-08DEC10-4/11

Basic Yield Monitor Components

To determine instantaneous crop yield, a farmer must know four things: the grain flow rate through the combine's clean grain system, moisture content of the grain, the combine's travel speed, and the cutting width of the header. The grain flow rate is measured within the combine near the grain tank. The flow rate is measured in units of volume or mass per unit time (bu/sec or lb/sec). Travel speed can be measured in a number of different ways, which we will discuss, and has units of distance per unit time (mi/hr or ft/sec). Cutting width can be measured (in feet, inches, or number of rows), but is often "eyeballed" by the combine operator. If travel speed and cutting width are known, the area harvested per unit of time can be calculated. If the volume or mass of crop harvested per unit time and the area harvested per unit time are known, then the yield can be determined.

The following components are found in most common instantaneous grain yield monitoring systems. These components work together to measure the necessary flow and working rates and to calculate, display, and record crop yields:

- Grain flow sensor

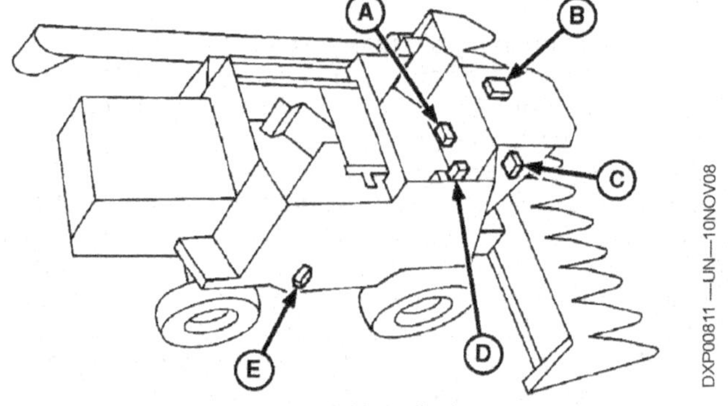

Instantaneous grain yield monitoring system

A—Moisture Sensor
B—GPS Antenna
C—Yield Monitor Console
D—Grain Flow Sensor
E—Ground Speed Sensor

- Grain moisture sensor
- Ground speed sensor
- Header position switch
- Display/processor console

Continued on next page

MM16633,000257C -19-08DEC10-5/11

Grain Flow Sensors

Methods of measuring grain flow vary, but a number
of popular yield monitoring systems use flow sensors
mounted in the path of clean grain flow. The grain flow
sensor is typically mounted at the top of the clean grain
elevator or conveyor, but in instances may be located in
other areas along the clean grain elevator.

A—Clean Grain Elevator   C—Loading Auger
B—Mass Flow Sensor

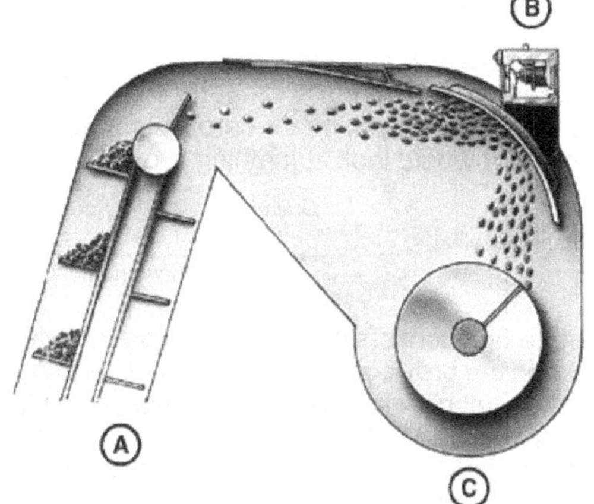

Impact-type mass flow sensor

MM16633,000257C -19-08DEC10-6/11

Grain Moisture Sensors

Grains are complex mixtures of compounds that include
proteins, starches, water, and oils. Grain quality, as it
relates to these compounds, is of increasing importance in
the marketplace. However, at harvest time, the farmer is
usually most concerned with only two grain components:
dry matter and water. Grain moisture content will affect
the timing of harvest. Moisture content affects the amount
of grain damage that will occur during harvest and how
grain must be handled and stored following harvest. Of
great importance is the effect of moisture content on
grain weight and volume. The buyer demands moisture
contents within a range that will permit storage and
handling with minimal losses.

Grain moisture content also affects the farmer's ability
to compare crop performance within and among fields.
Moisture contents can vary widely within a field and will
certainly vary over time. It is necessary to record moisture
content at the time of harvest so that all yield data can be
converted to a standard value. For corn, the standard
moisture content is 15.5% wet basis (weight of water
divided by the weight of water plus dry matter). Most yield
monitoring systems include some means of measuring
grain moisture content automatically, on-the-go. This
allows each yield data point to have an associated
moisture content value.

Capacitance-Type Sensor

The moisture sensor is generally located in the combine's
clean grain conveying system near the grain flow sensor.

A capacitance-type grain moisture sensor

A—Capacitance Sensor

A capacitance-type sensor is most often used to measure
grain moisture content. Capacitors accumulate and hold
an electrical charge on metal plates separated by a
dielectric. (A dielectric is a material that does not conduct
electricity and can contain an electrical field.) The sensor
measures the dielectric properties of the grain that flows
between the metal plates. The higher the moisture content
of the grain, the higher the dielectric constant. Therefore,
this measurement indicates the moisture content of the
grain.

Continued on next page   MM16633,000257C -19-08DEC10-7/11

Ground Speed Sensors

If the ground speed and grain flow rate are measured and the cutting width is known, instantaneous crop yield can be determined. The calculation for crop yield performed by a yield monitor would look something like the following:

$$\text{Instantaneous yield} = \frac{\text{Grain flow rate} \times \text{Unit conversion factor}}{\text{Cutting width} \times \text{Ground speed}}$$

An appropriate unit conversion factor must be included in the equation on the previous page. The reason is to permit grain flow rate, cutting width, and ground speed to be entered in convenient units such as pounds per second, feet, and miles per hour and give the result, instantaneous yield, in standard units such as bushels per acre.

Header Position Sensor

Some yield monitoring systems rely on a header position sensor to control the calculation of harvested acreage. When the sensor detects the header in the raised position, area counting is suspended, even when the combine is in motion and all systems are operating. When the sensor detects that the header has been lowered to a reasonable cutting height, area counting is resumed. The sensitivity of the sensor can be adjusted to permit header height changes that are necessary during harvest without shutting off the area counting process. This feature permits the combine to turn at field endrows and cross waterways and other non-crop areas without including the area covered in the yield monitor's harvested acreage calculations

Some yield monitoring systems include software that allows the operator to specify an operation delay to

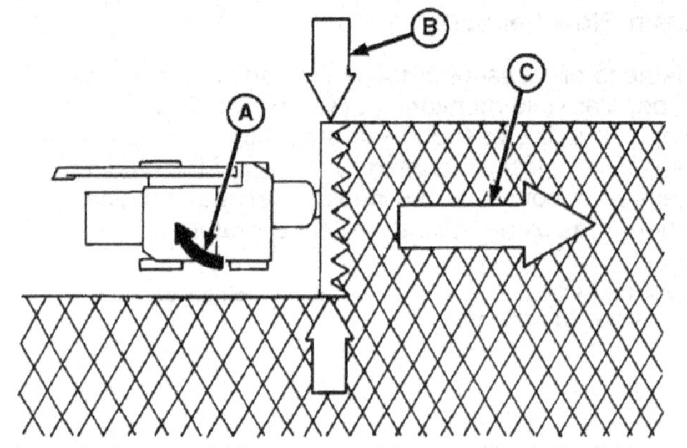

Instantaneous yield measurement by a combine in the field

A—Grain Flow into Tank
B—Cutting Width
C—Travel Speed

account for the time required for grain to move from the header to the grain flow sensor. Some systems include a start of pass delay to permit the initial flow of grain into the combine to be ignored in yield monitoring computations. The initial flow of grain past the flow sensor is often not indicative of actual yield; a characteristic "ramping" occurs before steady state grain flow conditions are achieved. On the other hand, an end of pass delay permits the grain flow that occurs after the header is lifted and the acre counting process is suspended to be included in yield monitoring computations.

Continued on next page

MM16633,000257C -19-08DEC10-8/11

Display Console

The monitor console or display unit is mounted in the
combine cab within easy view of the operator. The
console connects to all of the sensors that supply
information needed to calculate grain yield. In addition
to sensor inputs, the display console receives inputs
from the combine operator. This permits the operator to
provide data for which no sensor is installed (width of cut,
for example) or field or load information to permit tagging
or referencing of the yield and moisture data that is being
collected. For instance, the data collected from a field
could all be tagged with, or referenced by, a user-supplied
name such as "Field A" or "North Forty." Each load could
be tagged or referenced in a similar manner. Display
consoles include a keypad for information input and
a visual display. Information that can be entered or
displayed can include the following:

Operator-Supplied Information

- Field name
- Load name or number
- Cutting width

Sensed/Calculated Information

- Crop moisture content
- Instantaneous yield
- Average yield
- Area harvested
- Travel speed
- Quality of DGPS signal reception

Yield monitor display console

Close-up view of a yield monitor display console

Continued on next page   MM16633,000257C -19-08DEC10-9/11

Calibration

The types of calibration that are required by yield monitoring systems vary by monitor type. However, regardless of the type of monitor, yields are not measured directly. Instead, measurements of force, displacement, or volume, speed of material flow, crop moisture content, harvester travel speed, and working width are combined to produce an estimate of crop yield. Crop yield is a derived, or calculated, value. Calibration is performed to ensure that sensor data and operator input are properly used by a monitor to produce a final output in units of bushels per acre. In addition, force-and displacement-sensing grain flow meters must be calibrated to nullify, or cancel, the effects of machine vibration on yield readings. In some cases, flow measuring devices are so sensitive that vibrations caused by the movement of machine components such as the straw walkers and cleaning shoe will produce errors in yield data if not accounted for.

Most calibration processes involve comparing the weights of several loads estimated by a yield monitor with those measured on a separate set of scales. During calibration, the farmer harvests a series of full or partial grain tank loads. The loads are transferred to a wagon and each load is weighed on a separate set of scales and the data recorded. The farmer can then enter the actual weights into the yield monitor. Software within the yield monitor console then computes a set of calibration curves. An example of a calibration curve is provided below. The curves are fitted to the yield monitor data in such a way that the difference between the corrected weights calculated by the yield monitor and the actual measured weights is minimized.

If harvesting conditions remain similar to those that existed at the time of calibration, yield monitors can produce very high accuracies. If actual harvesting conditions (grain flow rates or moisture content, for example) vary a great deal from the conditions at calibration, yield estimate

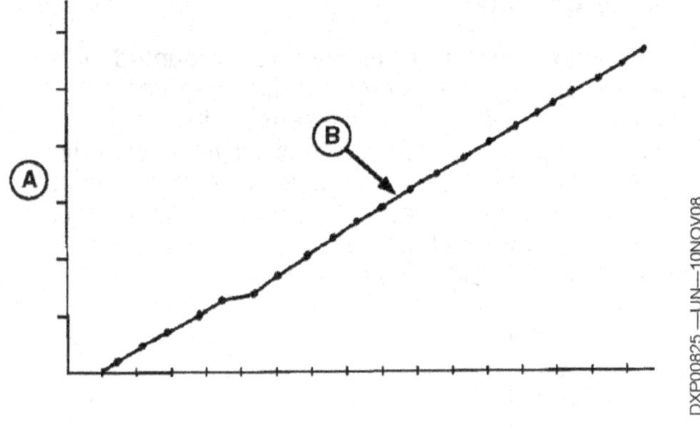

Simulated yield monitor calibration curve

A—Grain Flow

B—Simulated Yield Monitor Calibration Curve

accuracies can suffer. It is for this reason that it is wise to perform the calibration process over a range of grain flow rates. This requires the farmer to harvest at different speeds and cutting widths during calibration and to weigh a number of loads of grain. If changing environmental conditions over the course of a harvest season cause crop condition to change significantly, recalibration might become necessary, even if a farmer harvests only a single type of crop.

It is advisable to compare moisture content estimates from the yield monitor's moisture sensor with estimates from samples tested by a reliable moisture tester. Errors in measuring moisture content will lead to errors in measuring yields. Some yield monitoring systems permit the operator to adjust the readings from the moisture sensor to make them agree with the readings from a reference moisture tester.

Continued on next page

MM16633,000257C -19-08DEC10-10/11

Yield Data Collection

The components and procedures we've discussed so far are all a farmer needs to collect and record yield data. With these basic components, an operator can instantaneously observe yield, speed, moisture content, and so on as these data "flash" on the monitor console. The "on-the-go" sensors collect data every 1, 2, 3, or 5 seconds, depending on what rate the operator chooses. The console also summarizes the collected data. This summary provides an average yield and moisture content for larger areas. An operator may also view summary data on the console display.

The monitor can store an extensive record of crop production information including yield values for the season, farm, field, and load. With a data storage device in the monitor console, the operator can transfer monitor data onto the card. Having transferred collected data to the PC card, the user simply removes the card from the monitor console. The user then inserts the card into a PC data card reader on a desktop computer or into a PC card slot on a laptop or notebook computer. With the proper software, the operator can transfer data into a file that can later be accessed for printing summary yield data tables.

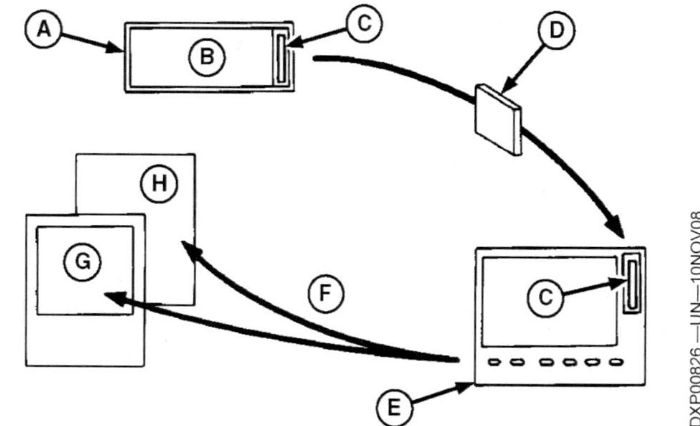

Data transfer within a yield monitoring and mapping system

A—Combine Console
B—Yield Monitor
C—Card Slot
D—PC Data Card (Physical Transfer)
E—Laptop/Farm Computer
F—(Via Color Printer for Color Maps)
G—Yield Maps
H—Yield Tables

MM16633,000257C -19-08DEC10-11/11

Yield Mapping

To produce yield maps that are based on sets of instantaneous yield data points, farmers need one more thing during harvest — a means of determining and recording the combine's in-field location for each yield data point. Each yield data point must correspond to a particular location on the earth's surface. The process of associating data such as crop yield values with geographic coordinate data is known as georeferencing. The most common georeferencing method for establishing and recording the locations of sites within a field is through the use of a global positioning system (GPS) receiver. To calculate a relatively accurate combine location estimate, the user needs some form or source of differential correction (DC) for the positioning system. Since yield monitoring is a data logging process and since the data can be processed after it is collected, it is not necessary to have a real-time source of differential corrections. However, most farmers have chosen to use real-time differential corrections because GPS hardware is typically used for a number of operations around the farm, some of which require high-accuracy position estimates. [Note — The GPS and DC receivers are shown as two separate items (antennas), but most suppliers of DGPS (differential GPS) equipment and services put both in a single package, with various antenna types and requirements.]

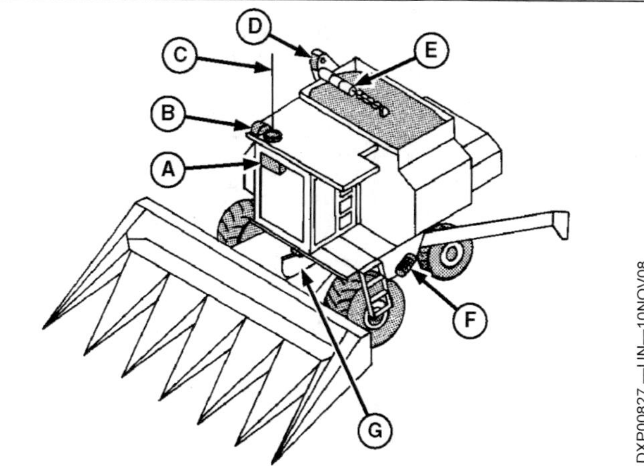

A combine equipped to collect yield data for mapping

A—Yield Monitor Console
B—GPS Receiver
C—DC* Receiver
D—Grain Flow Sensor**
E—Grain Moisture Sensor**
F—True Ground Speed Sensor
G—Header Position Sensor**

Continued on next page

OUO1023,00040C4 -19-23OCT15-1/4

If equipped with a GPS receiver, a yield monitoring system records the location of all yield data directly to a PC card. The card must remain in the yield monitor slot during harvesting since the internal memory of most yield monitors is not sufficient to store all the point data for whole fields. The farmer will use the georeferenced yield data later to generate yield maps. The procedure for generating maps resembles the process for generating summary tables, requiring the farmer to transfer data from the yield monitor to a computer via a PC card. Note that site-specific yield monitoring produces significant quantities of data. For example, let's re-examine the combine mentioned earlier. It has a 20-foot wide header and travels in the field at four miles per hour. If the farmer chooses to record yield data at one-second intervals, there will be more than 370 data "points" collected per acre.

Users need special computer software to generate yield maps. PC cards store yield data in binary image files. Binary files store digital data very efficiently (computer storage space requirements are minimized). However, binary image files must be converted to standard text files in order for most yield mapping software to process the data. Standard text files can be viewed and understood by the user while binary image files cannot. The text files can also be imported with ease into spreadsheet programs. When creating yield maps from text files, it is recommended to use a computer with significant memory, data storage capacity, and processor and display capabilities. A number of examples of grain yield maps produced by commercially-available software packages appear on the next page.

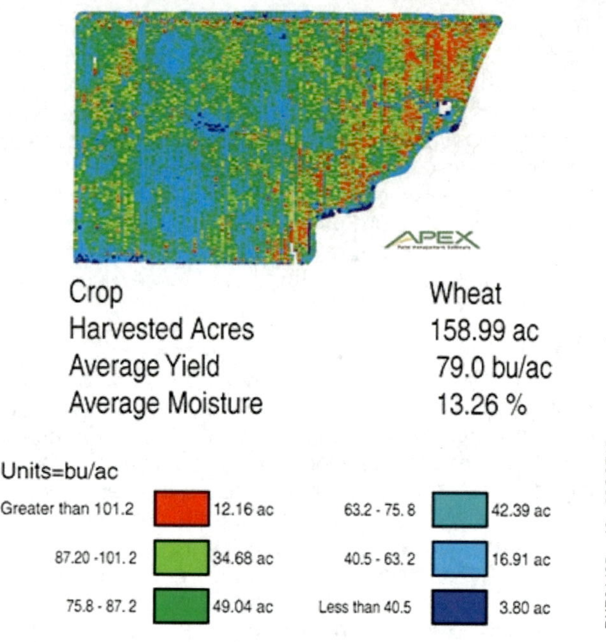

A yield map showing data collected at one-second interval

Continued on next page

OUO1023,00040C4 -19-23OCT15-2/4

What Yield Maps Can Reveal

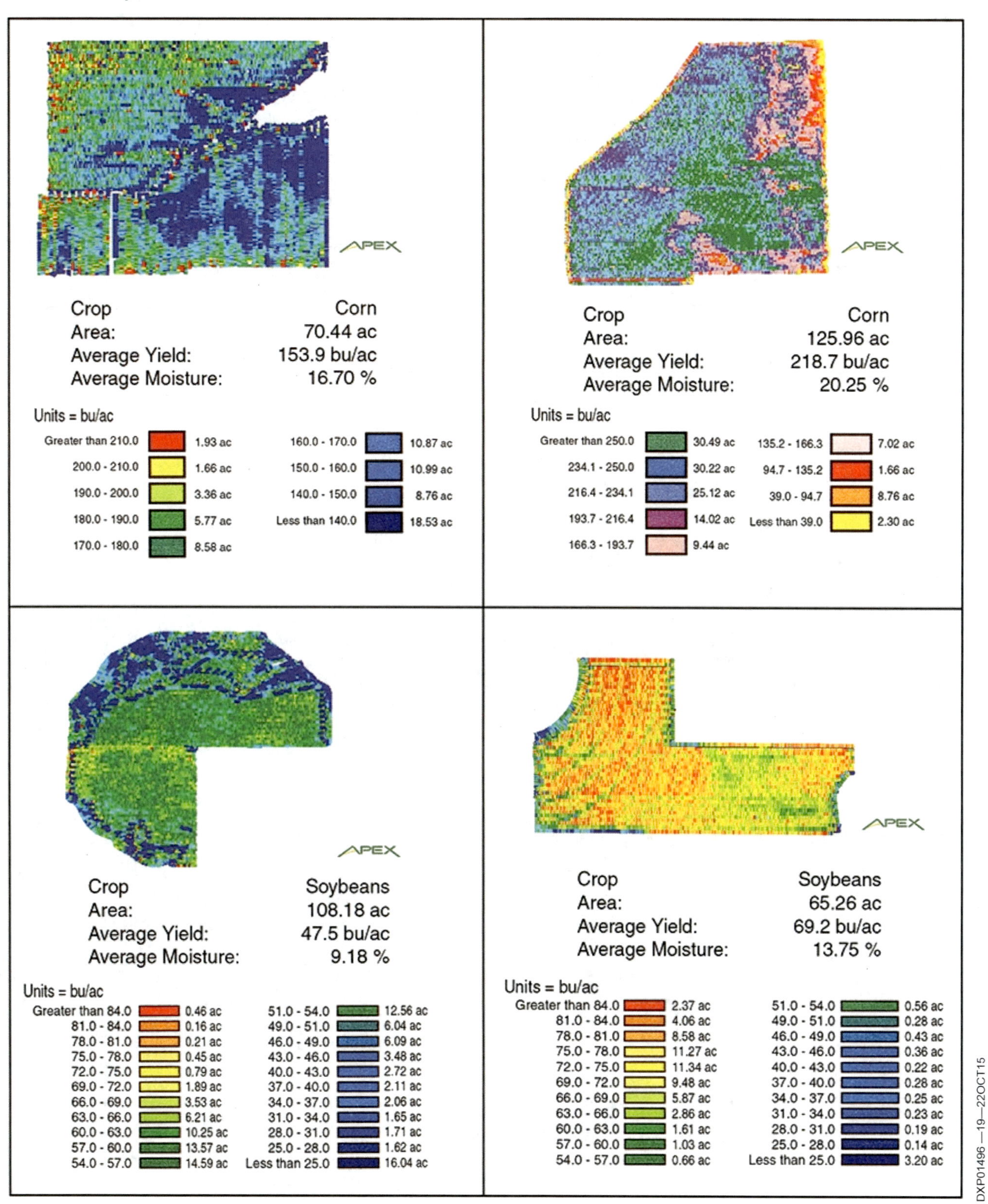

Examples of grain yield maps

Continued on next page

OUO1023,00040C4 -19-23OCT15-3/4

DXP01496 —19—22OCT15

102615
PN=119

Farmers know that yields vary within a field — they see it as material enters the combine header and hear and feel it as the crop is threshed. They may even have some idea of how much yields vary. But the new generation of instantaneous yield monitors can put actual numbers to their speculations about yield variability — which might be enough reason to use them. Furthermore, with these numbers and the yield maps, farmers can investigate why the yield variations occur. They can begin to establish relationships between yield variability and yield limiting factors such as soil type differences, or problems associated with fertility, weed control, drainage, soil compaction, equipment malfunction, and so on. In fact, a number of farmers claim their yield monitoring system "paid for itself" in the first year of use based on what they discovered by studying yield maps.

Certainly not all first-time users are that fortunate. But yield monitors can also be extremely useful in other ways. For instance, they can be used to conduct "on-farm research" — trials that investigate hybrids, seeding or fertilizer application rates, pesticide types and rates, and so on. The information gathered through the analysis of yield data collected over a number of years can forever change the way a farmer deals with chemical and seed suppliers. Accurate, long-term yield information has even been credited with increasing the value of such monitored farmland at the time of sale.

OUO1023,00040C4 -19-23OCT15-4/4

Soil Sampling and Analysis

For years agronomists and soil scientists have encouraged farmers to sample and analyze the soil of their fields on a regular basis. Regular soil sampling is important for developing a successful fertility management program. However, in the past, soil sampling was often overlooked and fertilizer was frequently over-applied to insure nutrient levels were adequate. With the recent interest in precision farming, more and more farmers are finally beginning to adopt regular soil sampling programs.

Methods of Soil Sampling and Analysis

Soil Sampling Instruction Sheet

1. To take a soil sample you need a sampling tube, auger or spade, and a clean plastic pail. Get sample containers from the lab where you are sending the samples for analysis.

2. Sample different soils, or areas treated differently in the past. Get equal-sized cores or slices from 15 or more places in each sampling area using probe, auger, or spade. Do not mix light- and dark-colored soils together.

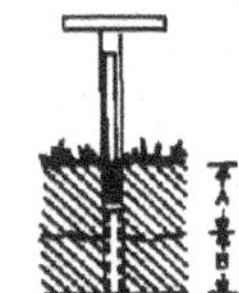

3. Take samples for P, K, and lime from the top 0-8 inches in tillage systems such as moldboard plow, chisel, and/or disk. Take samples for P and K from the top 0-8 inches in no-till fields and forage stands. Take samples from the top 0-4 inches for lime only in no-till fields. No-till fields which will be plowed periodically should be sampled to plow depth.

4. Place cores or slices in a clean plastic pail. Mix them together thoroughly, breaking up the cores or slices. If soil is muddy, dry it before mixing. If soil crumbles easily, dry after mixing.

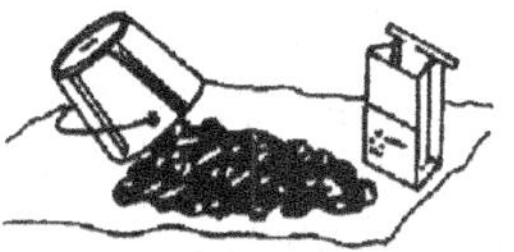

5. Spread mixture out on clean paper to dry. Do not heat in the oven or on the stove. Do not dry in places where fertilizer or manure may get in the sample.

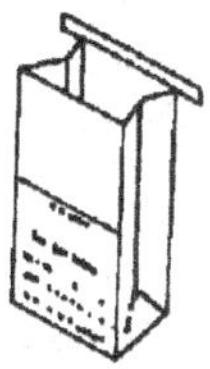

6. Fill the sample bag to the line with air-dry soil. Discard the rest. Label and number the sample container.

7. Identify the sample and record the cropping and fertilizer information requested using a field and cropping information sheet provided by the lab doing the analysis.

8. Draw a field sketch or farm map on a separate sheet and keep it in your files for your records, and to assist in developing your management plan.

Soil sampling process

In the past, farmers estimated the conditions of an entire field by averaging the results from analysis of soil samples randomly gathered around the acreage. Then the entire field was treated based on the average analysis. This approach of treating a whole field on the average made fertilizer recommendations and applications very simple. Only one rate of fertilizer was applied. With new precision farming technologies that allow changing fertilizer rates on-the-go, fertilizer is only applied as needed within each management unit in a field. This change in application methods has shifted the goal of soil sampling from measuring the whole field average to measuring the variability of soil properties throughout the field. The two most common methods that accomplish this are:

- Grid sampling
- Soil type sampling

DXP00841 —19—29SEP10

Continued on next page

MM16633,000257F -19-08DEC10-1/4

Grid Sampling

Grid sampling involves dividing a field into square or rectangular sections of several acres or less in size. The farmer gathers soil samples from each section and sends them to a laboratory for analysis. The objective of this approach is to better estimate the need for soil nutrients on a scale smaller than the entire field.

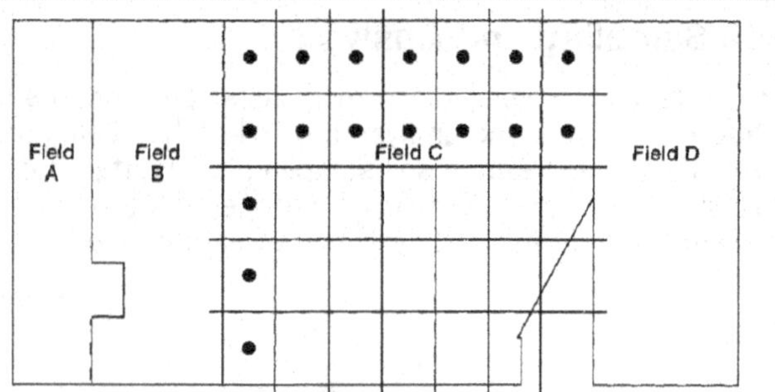

Dividing the field into small cells by placing a square grid over a field map is the first step in grid sampling

MM16633,000257F -19-08DEC10-2/4

At this time, researchers are still investigating different methods to determine the "best" way to sample each cell. Not many commercial soil testing companies use the grid cell method because it requires the person sampling to entirely cover each cell instead of just grid cell centers.

A—Regular
B—Staggered Start
C—Random Start
D—Systematic Unaligned
E—Random Cluster
F—Simple Random

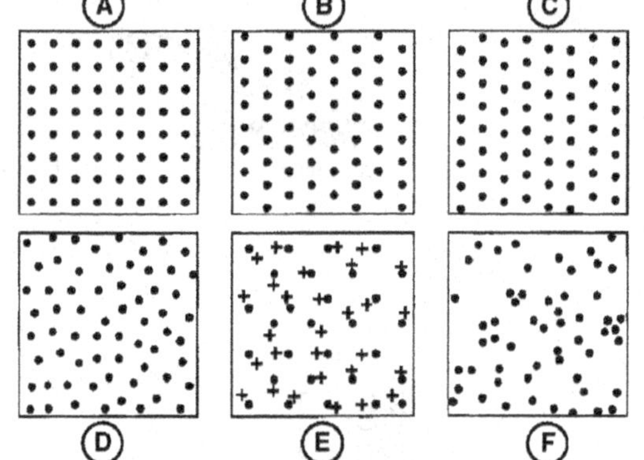

Alternative sampling patterns for grid cell sampling

Continued on next page

MM16633,000257F -19-08DEC10-3/4

Soil Type Sampling

An alternative to sampling on a square grid is to sample sections of the field that have similar soil types. The farmer employs the same sampling procedures. However, instead of blindly using a uniform grid, he or she uses tools such as soil survey maps to select sampling locations. Several samples are collected from areas having a particular soil type, then combined to represent that soil type. This is done for each soil type of interest within a field. This method results in samples being taken from irregularly-spaced sites around a field.

Summary

Soil testing is an important component of any crop production system. As farmers turn to precision farming methods and site specific crop management (SSCM), they find that a comprehensive soil sampling and analysis program is even more critical. In most cases, the measurement methods and the soil properties analyzed are the same as agronomists have recommended for many years; N, P, K, pH, etc. Yet today's precision farmer can use new methods and technologies to improve the precision and reliability of the soil sampling programs. First, positioning systems can pinpoint sampling locations, ensuring an accurate map of the soil conditions if sampling density is sufficient. Second, because farmers are beginning to recognize the value of soil tests and the variability occurring within each field, they gather more samples per acre and tend to sample more frequently. In addition, farmers can now use computers to record the results in digital maps, providing essential reference data for soil management. These three factors can help

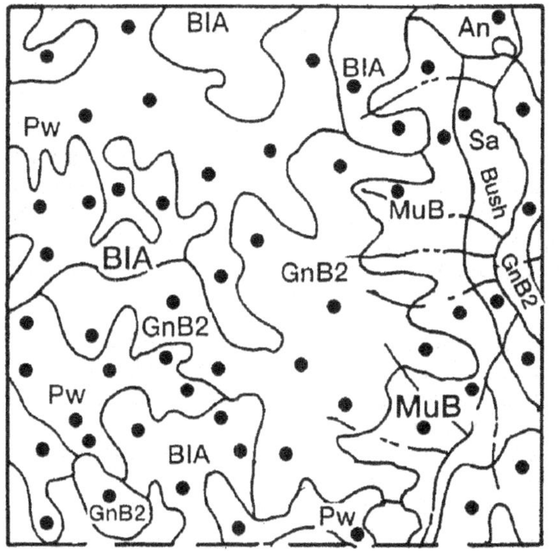

Sample locations used for soil type sampling

reveal the differences in soil properties within a field so the farmer can knowledgeably manage seeding, fertilizer, or herbicide rates based on the soil condition. After all, the concept of varying these inputs assumes you have measured some field variations in order to recommend changes in rates. Finally, measuring variability in soil properties is becoming automated as a growing number of on-the-go systems for measuring organic matter content, soil moisture, electrical conductivity, and compaction are being developed for the precision farming market.

MM16633,000257F -19-08DEC10-4/4

Remote Sensing

Every time a farmer looks out over a cornfield we could say that he or she is "remotely sensing" the status of the growing crop. By looking at the color and shape of the leaves one can make a visual assessment of whether the crop is doing well or is under some sort of stress from drought, nutrient deficiency, or pest infestation.

Remote sensing has gained a lot of interest as a potential management tool for precision farmers. Images from satellites and aerial platforms may allow the farmer to quickly view crops on his or her entire farm and decide which areas need further attention without leaving the comfort of the farm office. The viewing part, as we shall see, is easy since technology is available to view objects at great distances from air and space. However, interpreting what we can see and making decisions based on such interpretation is still quite complicated. Even though remote sensing has been used in agricultural applications since the 1930s, the practice is still undergoing development as a management tool for crop production.

Remote sensing (RS) has been defined in many ways. In general, RS is a group of techniques for collecting

information about an object or an area without being in physical contact with that object or area. The distances separating the sensor from the object or area being sensed can range from a few meters to thousands of kilometers. The most common methods of collecting the information include using either aircraft-based or satellite-based sensors. A familiar example of remotely sensed data is the weather maps commonly seen on television news showing cloud cover and precipitation around the country. These cloud data are usually collected from geostationary weather satellites or from satellites orbiting the Earth. Rainfall intensity data are collected from radar tower installations located around the country.

Many different sensors for both aerial and satellite imaging are used for these remote sensing applications.

Remote sensing of crops can be an attractive alternative to the traditional methods of field scouting because of the capability of covering large areas rapidly and repeatedly. Remote sensing provides the farmer with a means of identifying potential problems before the damage becomes irreversible in terms of crop yield or quality.

MM16633,0002582 -19-04OCT10-1/1

Computers and Geographic Information Systems

Precision farming activities like yield monitoring, crop scouting, or grid soil sampling provide data about the variation in crop and soil conditions throughout the field. This data must then be processed into maps to provide serviceable information. This section describes the basic computer hardware and software for analyzing precision farming data and converting it into useful information which can be displayed on colored maps.

One of the drawbacks of precision farming is that an enormous amount of data is collected during activities like yield monitoring and soil testing. Computer software programs that can easily store, manipulate and display this data are important tools for a precision farming operation. The software can range in complexity from a simple data displaying package, or mapping software, to sophisticated packages capable of analyzing multiple layers of data.

A geographic information system (GIS) is really just a type of data base system. A data base system provides for the input, storage and retrieval of data. The actual set of data that is stored is called the data base. In the case of a GIS, the data base contains geographic data. Geographic data contain not only the attribute being reported but also the spatial location of the attribute. An example of an attribute would be the value of yield produced by a particular spot in a field. The location is usually specified by a common coordinate system. Latitude and longitude is an example coordinate system used for referencing yield values. The following table shows geographic data from a yield monitor referenced with latitude and longitude coordinates.

Latitude degrees	Latitude degrees	Corn Yield Bu/acre	Corn Moisture percent
41.8048984	-91.8310704	185.1	19.0
41.8048862	-91.8310705	185.4	18.5
41.8048750	-91.8310718	184.2	18.4
41.8048632	-91.8310743	183.9	18.9
41.8048529	-91.8310752	182.1	19.0
41.8048420	-91.8310765	181.0	18.5
41.8048317	-91.8310718	179.9	18.8
. .	.	.	
. .	.	.	
. .	.	.	
41.8047288	-91.8310645	182.7	18.7

Geographic data from a yield monitor

Geographic data is sometimes referred to as spatial data. For our purposes, both terms can be used interchangeably. When using geographic or spatial data,

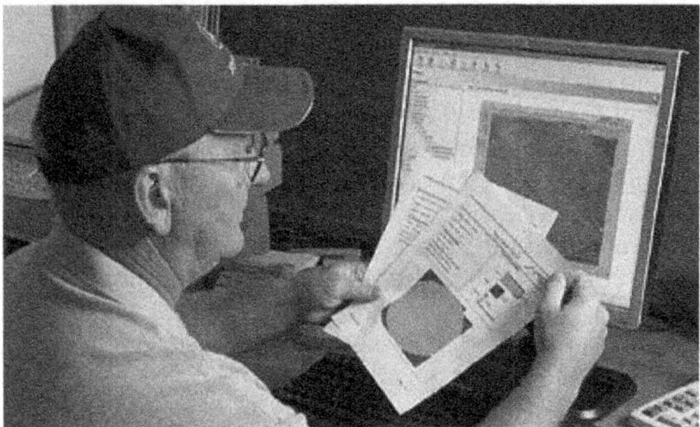

In precision farming, a computer is necessary to analyze the large amount of data that is generated

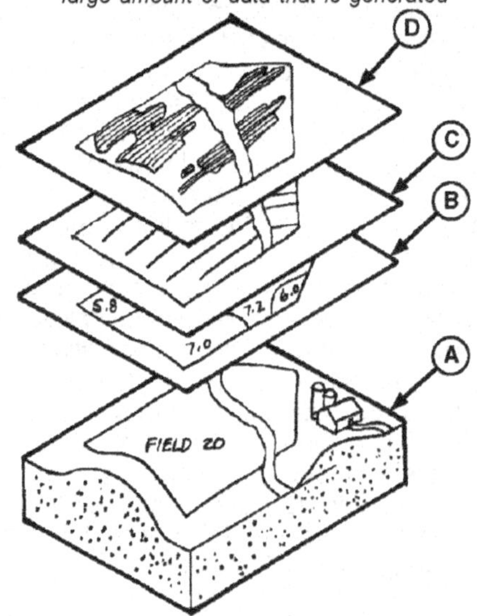

A GIS can display and analyze data as multiple layers of the same field. Each layer may represent one of the following: field boundaries, soil pH, drainage tiles, yield values, etc.

A—Boundary Layer
B—Soil pH Layer
C—Drainage Tile Layer
D—Yield Layer

what it is and where it is are important. Spatial data that pertains to a location on the earth's surface is called georeferenced data. A GIS used in precision farming georeferences data by attaching a physical location in a field with that location's values. The power of a GIS is its ability to link more than one data value, such as yield, nutrients, or soil type with the same georeferenced coordinate. Then, these different data values can be displayed in different layers overlaying the same field.

MM16633,0002585 -19-09DEC10-1/1

Variable Rate Technologies

Farmers can spend a lot of time and money collecting and analyzing the data described earlier and get no return on investment unless something useful is done with the information. This section describes methods of making use of the site-specific data to implement variable-rate application (VRA) of cropping inputs such as seed, fertilizer, and pesticides. The equipment to perform VRA is commonly called variable-rate technology or VRT. VRA is only one management approach for addressing the within-field, spatial variability described previously. Site specific crop management (SSCM) is a term that more broadly describes the use of variability in soil and crop parameters to make decisions on the precise application of production inputs. In this light, VRA can be considered one method of implementing SSCM.

Components of All Variable-Rate Applicators

The technologies for implementing VRA in sensor-based and map-based systems share most major components. Therefore, we will discuss the technologies in a general manner, then show how they are used in the specific areas. VRA automatic control systems have three major components:

- Sensors — positioning, pressure/flow, ground speed
- Controllers
- Actuators

Sensors for VRA

Positioning systems are listed under sensors since they provide position inputs to the controller. DGPS is the most common positioning system used for VRA. The DGPS used here must be real-time DGPS as described earlier.

Soil and plant sensors for the following purposes have been developed for VRA:

- Soil organic matter content (SOM)
- Soil moisture content
- Light reflectance of crops and weeds
- Soil nutrient level

Variable-Rate Controllers

Controllers are the devices that change the application rate of products being applied on-the-go. Controllers use microprocessors to "read" sensor inputs and calculate a product output rate based on stored algorithms. In the context of VRA, an algorithm is a formula or "recipe" that relates inputs from sensors or maps to an output that controls a pump or valve on a machine.

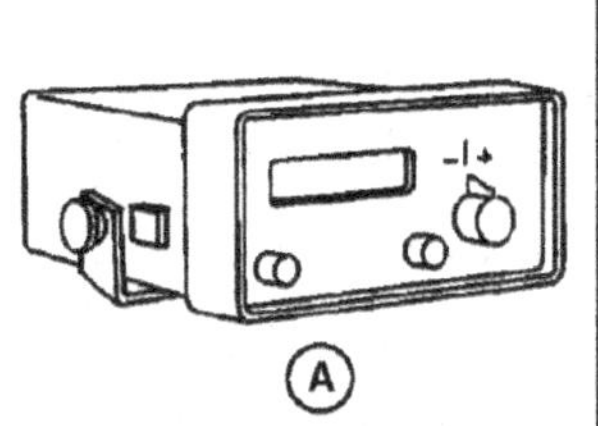
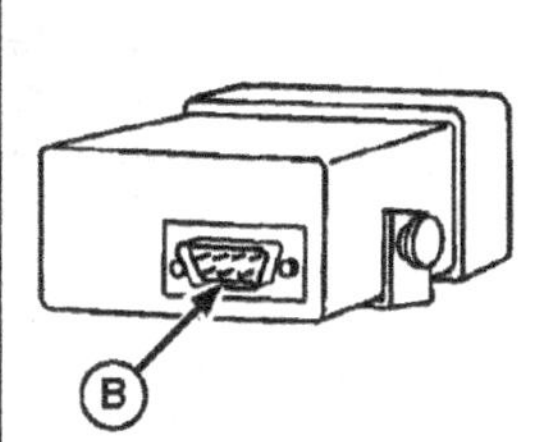

VRA controller with a serial port interface

A—VRA Controller   B—Serial Port on Back

Note that not all equipment controllers are designed for sensor-based or map-based VRA. Many equipment controllers only allow the operator to manually change application rates through manual switches.

Actuators

Actuators are devices that respond to signals from controllers to regulate the amount of material applied to farm fields. The actuator's response might be to extend or retract, rotate a shaft, open or close a gate, or cause speed to change. Actuators are designed to respond to electrical, pneumatic, or hydraulic signals that originate from a controller. An actuator can be used to change the position of a valve that regulates fluid flow rate or pressure.

An actuator can be used to change the position of a sliding gate to regulate the flow of granular material onto a conveyor. An actuator can change the output of a pump such as the one illustrated below, commonly used for pumping liquid fertilizer solutions.

Technologies for Variable Rate Applications

VRA systems can be categorized by the type of product that is applied:

- Seeds
- Dry chemicals (granular fertilizer, granular pesticides, limestone)
- Liquid chemicals (liquid fertilizer, liquid pesticides)

The three major types of crop production inputs applied in spatially-variable or variable-rate operations have been: fertilizer and limestone, pesticides, and seed. The systems incorporate features such as direct injection and on-the-go blending to permit variable-rate applications. The following provides descriptions of examples of VRA systems.

Continued on next page   MM16633,0002587 -19-09DEC10-1/6

Variable Seeding-Rate Planters

Planters or drills can be made into variable rate seeders by independently adjusting the speed of the seed metering drive. Conventional row-crop planters utilize a ground-driven wheel to drive the seed metering system. The ground drive and metering systems are connected by a chain drive to provide a constant ratio of ground wheel speed to seed disk or seed pickup speed. This keeps the seed spacing within each row constant over a range of travel speeds.

For row-crop planters, the seeding or planting rate is determined by the spacing between planter units and the spacing between seeds within each row. Seeding rate is varied by changing seed spacing, since row spacing is fixed. Variable rate seeding is accomplished by separating or disconnecting a planter's seed metering system from its ground drive wheel. By uncoupling the drive wheel and metering system and using another drive mechanism, like the hydraulic motor illustrated, planting rate can be varied on-the-go.

Conventional planters use a ground drive system to control the seed metering device

A hydraulic motor used as a variable speed drive on the seed metering shaft

Continued on next page

MM16633,0002587 -19-09DEC10-2/6

Dry Chemical Applicators

Applicators used for variable-rate application of dry chemicals include both spinner spreaders and pneumatic applicators. Spinner spreaders have the ability to apply more than one product at a time via one or two spinning disks at the rear of the applicator and allow for quick application rate changes. A conveyor belt, or chain, transfers material from a hopper and feeds it onto the spinning disks (spinners).

The granules pass through an adjustable gate opening that controls the amount of material dropped onto the spinners. The dry chemical application rate can be adjusted by changing the gate opening or changing the speed of the conveyor belt.

VRA pneumatic applicators have centrally-located bins, or hoppers, and air tubes leading from a metering unit to the point of discharge. Variable rates of dry chemical are suspended in the air stream. The dry product is distributed by a deflector plate at each outlet along the length of the applicator's boom. Single or multiple products can be metered independently on-the-go with metering controls on each product bin.

Spinner spreaders are commonly used for applying granular fertilizers and limestone

Dry product is distributed by a deflector plate at an outlet on the applicator boom

Continued on next page

MM16633,0002587 -19-09DEC10-3/6

Liquid Chemical Applicators

Variable-rate liquid applicators or field sprayers are designed to provide adjustable product output rate (volume per unit time). But, application rate (volume per unit area) is affected by both product output rate and applicator travel speed. So, to maintain a pre-set application rate, travel speed is monitored and output rate is adjusted to compensate for changes in ground speed. Fine adjustments in application rate have been accomplished by changing system operating pressure. Coarse adjustments have been accomplished by changing nozzles (orifice size) or travel speed. Now, new technologies provide new options for adjusting liquid chemical output and application rates.

Spray controllers using a flow control valve are either flow- or pressure-based. In other words, the measurement from either a flow meter or a pressure sensor is used to insure proper control valve adjustment. Flow rate through the nozzle is proportional to the square root of the pressure at the nozzle. However, some conventional nozzles are designed to operate over a very narrow range of pressures because changing pressure affects the spray pattern and droplet size distribution. Unwanted changes in spray pattern and/or droplet size distribution can cause spray coverage and drift problems. Newer air induction nozzles allow for a wider range of pressure. They maintain a consistent spray pattern and droplet size without the need to change the nozzle size for a variety of flow rates.

Sprayer system for liquid chemical application

Using changes in pressure to produce large changes in flow rate is difficult. Though, the use of a controller provides a selection from among combinations of nozzles, which can allow for a wide range of potential flow rates.

By switching among nozzles of different size on-the-go, discrete changes in flow rate can be achieved without having to vary system pressure and spray distribution pattern significantly.

MM16633,0002587 -19-09DEC10-4/6

An alternative to multiple-nozzle spraying systems has been developed through the use of modulated spraying nozzle control (MSNC), also known as pulse width modulation (PWM) technology. Fast-acting solenoids are used to rapidly turn on and off the flow of liquid to individual spray nozzles. The switching process takes place at a rate of 10 Hz (or 10 cycles per second). By modulating the position of a solenoid attached to a valve positioned in line with a spray nozzle, effective flow and application rates can be varied over a wide range. Both VRA and drift reduction can be accomplished by the same application system.

An alternative to varying sprayer flow rate or pressure is to change the amount of active chemical added to a constant rate flow stream. Varying the amount of chemical product injected into a carrier that is delivered at a constant rate is desirable because of the consistency in spray pattern and droplet size.

Illustration of a conventional nozzle equipped with a sole-noid-actuated control valve

A—Open Valve
B—Closed Valve
C—Conventional Nozzle Assembly
D—Solenoid

Continued on next page

MM16633,0002587 -19-09DEC10-5/6

Another fertilizer product that is handled as a liquid, but at high pressures with specialized equipment, is anhydrous ammonia. Liquid anhydrous ammonia is normally knifed into the soil so that evaporation loss is minimized.

Since the boiling point of NH_3 is -28°F, the vapor pressure of liquid ammonia is typically used to pump the liquid from a carrier tank. A regulator valve is also needed to control the flow as vapor pressure varies with amount and concentration of ammonia in the tank and with temperature. In a typical anhydrous ammonia application system, a thermal transfer unit is used to convert vapor to liquid for metering and application because measuring flow rates of NH_3 gas is difficult. Other techniques have been developed to convert ammonia vapor to liquid, but the need to have the product in a liquid state to ensure accurate metering applies across-the-board. Accurate metering is essential to permit anhydrous ammonia to be part of a farmer's variable-rate, precision farming system.

Applicator for applying anhydrous ammonia

DXP00926 —UN—10NOV08

MM16633,0002587 -19-09DEC10-6/6

Machinery Management

Introduction

Machinery management has increased in importance in today's farming operations because of its direct relation to the success of management in combining land, labor, and capital to return a satisfactory profit. Machinery costs include fixed (ownership) costs and variable (operating) costs.

Modern agricultural machinery must be skillfully managed for maximum profits

Continued on next page

MM16633,0002603 -19-09DEC10-1/2

PN=130

The importance of machinery in the total farming operation is indicated by the machinery costs in relation to the total costs. Typically, machinery costs overshadow all other crop production costs except land. Machinery costs often account for 50% of operating production costs. It is not unusual to find that the difference in profit from one farm to another is due solely to differences in the machinery selected and the way it is managed.

Typical Problems

The problems of machinery management are quite varied. Following are some typical examples of the more important problems. In each case the decision could mean the difference between a profit and a loss.

- How much machinery should be owned?
- What size equipment is needed?
- Should a custom operator be used?
- How often should machinery be traded?
- Should a machine be repaired or traded?

These are only five examples of the many important decisions related to owning and operating agricultural machinery.

For more information regarding machinery management refer to Machinery Management FBM17106NC.

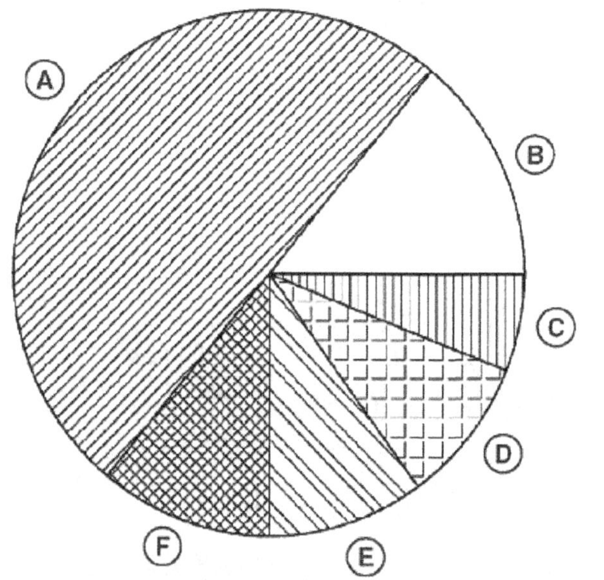

Excluding land costs, machinery costs overshadow all other costs. Because they change, current rates should be reviewed.

A— Machinery (50%)　　D— Seed (10%)
B— Other (13%)　　E— Fertilizer (10%)
C— Chemicals (6%)　　F— Labor (11%)

MM16633,0002603 -19-09DEC10-2/2

Measuring Machine Capacity

Estimating capacity is important in order to match machine size to available working time

The capacity of a machine is its rate of performance. Depending on the kind of machine, the performance or capacity will be measured in terms of acres per hour (hectares per hour), tons per hour (metric tons per hour), bushels per day, or hundredweight per hour (quintals per hour). It is important for efficient farm managers to understand how to estimate capacities of machines they plan to buy for future use.

Continued on next page

MM16633,0002604 -19-15OCT10-1/2

PN=132

Capacity Measuring Methods

The capacity of a machine is its rate of performance, usually reported in terms of quantity per time. Acres per hour (hectares per hour) is the most common measure of machine performance. Harvesting and processing machine performances may be measured as bushels or tons per hour. Bushels per hour is a rather inaccurate measure due to variations in moisture content and grain densities. In many cases the hundredweight (cwt) measure, which is 100 pounds, is used. In metric units, the common units include kilograms (kg), quintals (100 kg), and metric tons (1000 kg).

Capacity of harvesting machines, in some cases, requires measurements other than area per unit of time.

Three possible expressions of machine capacity include:

- Field capacity in acres or hectares per hour
- Material capacity in hundredweight (cwt) or kilograms per hour
- Throughput capacity in pounds, tons, kilograms, or metric tons per hour

Field capacity is the most commonly used measure of machine capacity.

Material capacity, once commonly referred to in bushels per hour, is now usually measured by hundredweight (cwt) or kilograms per hour. In many cases, tons per hour is used. Material capacity is simply a measure of material, such as silage or grain, harvested by a machine. But, in the case of harvesting machines such as combines, this is not a completely accurate expression of capacity, because it represents only the amount of grain or forage harvested. To indicate total amount of material passing through the machine, throughput capacity is used.

Throughput capacity is a special term used to compare machines, such as combines and potato harvesters, that separate undesirable material from the desirable material. In these cases, the weight of the material handled is the accurate capacity measure. Throughput, then, refers to the total weight of material processed in a given period of time, usually in terms of pounds (or kilograms) per hour. In the case of a combine, the pounds-per-hour (or kilograms-per-hour) throughput would include grain, chaff, straw, and any other material that enters the header. Because moisture affects the result, a moisture report should accompany the throughput capacity rating.

Let's consider a sample problem using these three different capacity measurements. It is determined that a combine with a 20-foot (6.1-meter) wide header is combining wheat at a speed of 300 feet (92 meters) per minute. In a 1-minute time period, 1,000 pounds (454 kilograms) of material enters the header. Of this amount, 500 pounds/min (227 kilograms/min) enters the grain tank and the remaining 500 pounds (227 kilograms) of crop residue is discharged through the combine.

Use the unit-factor method to determine the different capacities.

Field Capacity

Speed times width equals field capacity:

300 feet/minute x 60 minutes/1 hour x 20 feet x 1 acre/43,560 ft^2

Answer = 8.26 acres per hour

Units of measure in the numerator cancel the identical units of measure in the denominator. Apply the unit-factor method to the following equations:

Material Capacity

Pounds (kilograms) harvested per hour:

500 pounds/minute x 60 minutes/1 hour x 1 cwt/100 pounds

Answer = 300 cwt (hundredweight) per hour

This answer could also be expressed in tons per hour:

300 cwt/hour x 100 pounds/1 cwt x 1 ton/2,000 pounds

Answer = 15 tons per hour

Throughput Capacity

Total amount of material per hour:

1,000 lb/minute x 60 minutes/1 hour = 60,000 pounds per hour = 30 tons per hour

These examples are briefly covered here to familiarize you with the technical capacities. For most practical farm or ranch applications, acres per hour and tons per hour are the preferred ways of measuring capacity. Material capacity requires accurate measuring techniques and is easily affected by moisture, yields, and other factors such as platform cutting height.

All of these capacities are theoretical capacities, not effective capacities. Effective capacity brings in the factor of efficiency. After reviewing practical methods of measuring machine capacity, the factor of efficiency will be introduced.

MM16633,0002604 -19-15OCT10-2/2

Improving Field Efficiency

The ability to improve field efficiency is the next important step in developing machinery management skills. There are several important reasons why a machine may have a certain field efficiency. Some lost-time factors are built into the operation. Other lost-time factors can be eliminated by good planning and management. Typical factors causing lost time include:

- Unused capacity
- Filling procedures
- Unloading procedures
- Turning and field conditions
- Unclogging machines
- Making adjustments
- Breakdowns
- Servicing machines
- Rest stops
- Changing operators
- Checking machine performance
- Unmatched machine capacity

Pre-season inspection programs help to ensure that the equipment does not break down during the use season. Technicians are trained to spot problems before they occur. They inspect the equipment using checklists developed to identify components that are near the end of their useful life. Often this inspection includes checking critical pressures, measuring parts to see if they are within specifications, and sending fluid samples to laboratories.

Checking fluid condition is an important part of a machinery maintenance program. Engine oil samples test

A pre-season inspection improves field efficiency by help-ing to eliminate breakdowns

for various metals, antifreeze, dirt, and other elements. Multiple samples at regular intervals are required to establish a trend. The trace amounts of metals or other contaminants in the oil sample can identify an engine failure long before it becomes a catastrophic failure.

Coolant sampling is just as important as oil sampling. Proper coolant system maintenance can prevent major engine failure due to cavitation, or holes in the cylinder liners.

MM16633,0002605 -19-19NOV10-1/1

Matching Machine Size and Capacity

Proper machine size means having adequate capacity to complete critical field operations during the working time available.

Effective field capacity must be understood in order to fit machine size to the amount of work that has to be completed in a specific length of time. Effective field capacity is the actual rate of performance of land or crop processed in a given time, based on total field time.

Proper machine size means having adequate capacity to complete critical field operations during the working time available

MM16633,0002606 -19-15OCT10-1/1

Estimating Power Requirements

A major task facing modern farmers and ranchers is to match power units to the size and type of machines so all field operations can be carried out on time with a minimum of cost. It is important to match the power needs of the equipment to that supplied by the power unit.

If the tractor is oversized for implements, the costs will be excessive for the work done. If the implements selected are too large for the tractor, the quality or quantity of the work may be lessened or the tractor will be overloaded, usually causing expensive breakdowns.

Some factors to consider when selecting a power unit:

- Engine types
- Power ratings
- Soil resistance to machines
- Tractor sizes
- Converting power
- Matching implements
- Determining wheel slip
- Sizing for critical work

Although there is no official classification for the various tractor configurations currently available, the two common classifications are:

Power units must be carefully matched to equipment, regardless of size

- Two-wheel drive tractors
- Four-wheel drive tractors

In practice, two-wheel drive tractors are arranged so the driving wheels are the rear wheels, while four-wheel drive tractors use all four wheels as driving wheels.

MM16633,0002607 -19-15OCT10-1/1

Estimating Fixed Costs

One of the most important costs influencing profit in farming operations is the cost of owning and operating machinery. Machinery costs are one of the few costs that good management can minimize. Learning how to realistically estimate machinery costs will aid in reducing production costs.

MM16633,0002608 -19-15OCT10-1/1

Estimating Fuel Costs

Fuel and lubricants are true operating costs because fuel consumption is directly proportional to the amount of use.

The amount of energy needed per acre for performing operations such as disking or plowing is nearly constant, regardless of speed and the size of the tools and tractor being used.

To better understand how to estimate fuel needs for crop production, consider the following factors:

- Horsepower-hours (kilowatt-hours) of energy
- Fuel types
- Fuel consumption

All three factors are important considerations when estimating fuel needs.

Per-acre fuel requirements are generally constant regardless of the equipment size

Continued on next page

MM16633,0002609 -19-08DEC10-1/4

Horsepower-Hours of Energy

The amount of fuel needed per acre is in proportion to the amount of energy required. One way of expressing the amount of energy used is "horsepower-hours" (or kilowatt-hours). Reviewing material on power ratings will aid in understanding this concept.

Horsepower-hours (kilowatt-hours) may be used to measure the work of a tractor in the field. The amounts of energy required for typical farm operations are shown as horsepower-hours per acre units in Table 1.

One horsepower delivered for 1 hour is 1 horsepower-hour of energy. Likewise, 1 kilowatt delivered for 1 hour is 1 kilowatt-hour of energy. The type of power used in calculating horsepower-hours (and kilowatt-hours) is PTO power.

A 100-hp (75-kW) tractor operating at 80% of maximum power for 2 hours delivers 160 horsepower-hours (120 kilowatt-hours) of energy

Operation	Energy Required, PTO HP-Hrs per Acre	Gallons per Hour		
		Gasoline	Diesel	LP-Gas
Shred Stalks	10.5	1.00	0.72	1.20
Plow 8 Inches Deep	24.4	2.35	1.68	2.82
Heavy Offset Disk	13.8	1.33	0.95	1.60
Chisel Plow	16.0	1.54	1.10	1.85
Tandem Disk, Stalks	6.0	0.63	0.45	0.76
Tandem Disk, Chisel	7.2	0.77	0.55	0.92
Tandem Disk, Plow	9.4	0.91	0.62	1.09
Field Cultivate	8.0	0.84	0.60	1.01
Spring-Tooth Harrow	5.2	0.56	0.40	0.67
Spike-Tooth Harrow	3.4	0.42	0.30	0.50
Rod Weeder	4.0	0.42	0.30	0.50
Sweep Plow	8.7	0.84	0.60	1.01
Cultivate Row Crops	6.0	0.63	0.45	0.76
Rolling Cultivator	3.9	0.49	0.35	0.59
Rotary Hoe	2.8	0.35	0.25	0.42
Anhydrous Applicator	9.4	0.91	0.65	1.09
Planting Row Crops	6.7	0.70	0.50	0.84
No-Till Planter	3.9	0.49	0.35	0.59
Till Plant (With Sweep)	4.5	0.56	0.40	0.67
Grain Drill	4.7	0.49	0.35	0.59
Combine (Small Grains)	11.0	1.40	1.00	1.68
Combine, Beans	12.0	1.54	1.10	1.85
Combine, Corn and Grain Sorghum	17.6	2.24	1.60	2.69
Corn Picker	12.6	1.61	1.15	1.93
Mower (Cutterbar)	3.5	0.49	0.35	0.59
Mower-Conditioner	7.2	0.84	0.60	1.01
Swather	6.6	0.77	0.55	0.92
Rake, Single	2.5	0.35	0.25	0.42
Rake, Tandem	1.5	0.21	0.15	0.25

Continued on next page

MM16633,0002609 -19-08DEC10-2/4

Baler	5.0	0.63	0.45	0.76
Stack Wagon	6.0	0.70	0.50	0.84
Sprayer	1.0	0.14	0.10	0.17
Rotary Mower	9.6	1.12	0.80	1.34
Haul Small Grains	6.0	0.70	0.50	0.84
Grain Drying	84.0	8.40	6.00	10.08
Forage Harvester, Green Forage	12.4	1.33	0.95	1.60
Forage Harvester, Haylage	16.3	1.75	1.25	2.10
Forage Harvester, Corn Silage	46.7	5.05	3.60	6.05
Forage Blower, Green Forage	4.6	0.49	0.35	0.59
Forage Blower, Haylage	3.3	0.35	0.25	0.42
Forage Blower, Corn Silage	18.2	1.96	1.40	2.35
Forage Blower, High- Moisture Ear Corn	5.9	0.63	0.45	0.76
Haul Forage, Corn Silage	4.0	0.42	0.30	0.50

Table 1 — Average Energy and Fuel Requirements

Table 1 — Average Energy and Fuel Requirements

Fuel-Saving Tips

Operating cost will be reduced by cutting fuel consumption whenever possible. Some ideas to consider are:

- Reduce tillage, when possible. Additional tillage means additional use of fuel.
- Combine operations and operate the tractor at full load. This can save one-half gallon or more per acre because a tractor is more efficient when operating at full load. Two trips with a partial load require considerably more fuel than one trip fully loaded using combined operations.
- Shift to a higher gear and throttle back when pulling a light load. It takes one-third more fuel when pulling a light load with the governor control wide open than to gear up and throttle back.

⚠ **CAUTION: Be careful not to overload the tractor with a reduced engine speed and a higher gear.**

- Select tire size and match ballast for proper wheel slippage. For most operations, 10% to 15% wheel slippage gives the best combination of maximum power at the least fuel usage. Too much slippage can waste as much as 2 gallons per acre.
- Keep the tractor in top running condition. Follow a rigid maintenance schedule and have the dealer give the tractor or machine a periodic service check. A tractor that is operating poorly can waste up to 25% of the fuel.

MM16633,0002609 -19-08DEC10-3/4

- Follow recommended storage practices for fuel tanks. Fuel suppliers can provide some good suggestions on storage. Use rustproof tanks or keep them shaded and use a pressure cap that conforms to local regulations. Diesel fuel is less likely to evaporate than gasoline. For both diesel fuel and gasoline, leakage and contamination by dirt and water are possible. Fuel tank filters protect engines by removing contaminants.

A—Red Tank Exposed to Sun's Heat, 9.6 Gal (36 L) per Month Lost

B—White or Aluminum Tank Exposed to Sun's Heat, 6.0 Gal (23 L) per Month Lost

C—White Tank Protected by a Shade, 2.4 Gal (9 L) per Month Lost

D—White Shaded Tank Equipped with a Pressure Vacuum Relief Valve, 1.3 Gal (5 L) per Month Lost

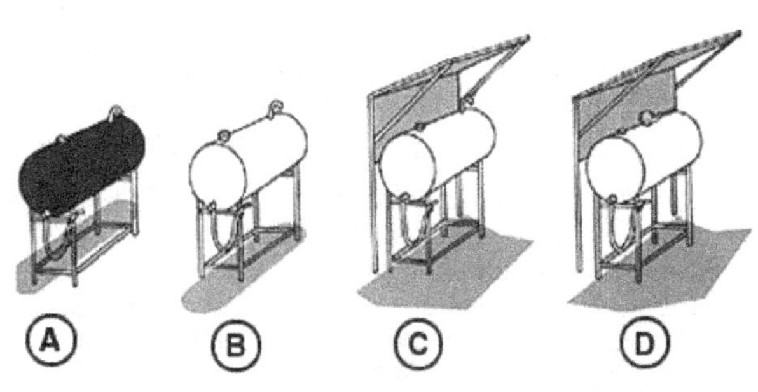

Summer evaporation losses from a 300-gallon (1,135 L) gasoline storage tank

MM16633,0002609 -19-08DEC10-4/4

Estimating Repair Costs

Repair costs, usually considered an operating cost, are another important part of machinery costs. The more a machine is used, the greater is its need for repairs.

Repair costs consist of all expenditures for parts and labor for repairs made in a shop or on the farm. In the case of older equipment, an estimate of deferred repair costs should be included. It is difficult to accurately predict repair costs for a specific machine. Repair costs will vary from one geographical section of the country to another because of differences in soils, crops, climate, and operators.

It is important to keep high-value operations going at full capacity

Machine	1/4 Accumulated Hours	1/4 Accumulated Cost	1/2 Accumulated Hours	1/2 Accumulated Cost	3/4 Accumulated Hours	3/4 Accumulated Cost	Full Life Accumulated Hours	Full Life Accumulated Cost	RF1	RF2
All Wheel Tractors	3,000	6.2%	6,000	25.0%	9,000	56.2%	12,000	100%	0.006944	2.0
Crawlers	4,000	5.0%	8,000	20.0%	12,000	45.0%	16,000	80%	0.003125	2.0
Self-Propelled Combines	750	2.2%	1,500	9.3%	2,250	21.9%	3,000	40%	0.039820	2.1
Cotton Harvesters	1,250	9.9%	2,500	24.4%	3,750	41.3%	5,000	60%	0.074044	1.3
Planters, Drills	375	4.1%	750	17.5%	1,125	41.0%	1,500	75%	0.320000	2.1
Moldboard Plows	500	8.3%	1,000	28.7%	1,500	59.6%	2,000	100%	0.287300	1.8
Chisel Plows, Mulch Tillers, Cultivators, Disk Harrows, etc.	500	10.1%	1,000	26.5%	1,500	46.8%	2,000	70%	0.265240	1.4
Mowers	500	14.2%	1,000	46.2%	1,500	92.0%	2,000	150%	0.461700	1.7
Large and Small Square Balers	625	6.2%	1,250	21.5%	1,875	44.7%	2,500	75%	0.144100	1.8
Large Round Balers	375	7.4%	750	25.9%	1,125	53.6%	1,500	90%	0.434000	1.8
Self-Propelled Forage Harvesters	1,000	3.1%	2,000	12.5%	3,000	28.1%	4,000	50%	0.031250	2.0
Self-Propelled Windrower	750	3.4%	1,500	13.7%	2,250	30.9%	3,000	55%	0.061100	2.0
Rakes	625	8.6%	1,250	22.7%	1,875	40.1%	2,250	60%	0.166300	1.4

Table 2 — Accumulated Repair Costs as a Percentage of List Price

Table 2 — Accumulated Repair Costs as a Percentage of List Price

Table 2 is a summary of repair costs for some of the more commonly used machines, and includes a listing of values for RF1 and RF2, which are repair cost parameters. Also listed are estimated repair costs for different amounts of machine use. Data in this table are based on 1998 Standards of the American Society of Agricultural Engineers. By plotting the formula it can be seen that repair costs increase with machine use at an increasing rate.

Information in Table 2 can be used in two meaningful ways:

- To help estimate total costs of a specific machine
- To compare known repair costs against what is considered average

Suppose a $50,000 list price tractor was purchased 5 years ago. Assume that $2,790 was spent over that time for repairs, which included all service shop costs, parts installed, and labor. The tractor hour meter shows 2,876 hours. Now, let's compare these repair costs to the estimate in Table 2. According to Table 2, a total of 2,876 hours is nearly one-fourth life. The table shows repair costs average 6.2% at that point.

$$0.062 \times \$50,000 = \$3,100$$

In our example, $2,790 was spent, which is $310 less than estimated from Table 2. Unless unusual repair costs are incurred in the next 124 hours of use, the tractor will have close to average repair costs or just under that amount.

Cost per hour = $2,790/2,876 hours = $0.97 per hour

The estimate from the table is:

$3,100/3,000 hrs = $1.03 per hour

Continued on next page

MM16633,000260A -19-11NOV10-1/2

Information in Table 2 can also be used for estimating repair costs for a specific machine. Suppose it is planned to do custom work with a $100,000 self-propelled combine and an estimate of repair costs is needed. Assume the combine will be used 1,500 hours, one-half its mechanical life. The combine operates at 7 acres per hour.

Information in Table 2 shows accumulated repair costs to be 9.3% of the new cost at 1,500 hours. Since 9.3% of $100,00 is $9,300, or $6.20 per hour, repair costs would be $0.89 per acre ($2.19 per hectare) at 7 acres (2.83 hectares) per hour.

Repair costs can usually be classified as an operating cost. Repairs are an important part of machinery management, because they are necessary to maintain a machine's reliability and keep it performing at maximum capacity.

Reliability expresses the amount of confidence placed in a machine to perform without unplanned time losses.

It is more important than ever to avoid or to be able to quickly repair machine breakdowns. Because of the larger machine size on farms today and the resulting increased productivity, good management of repairs and the resulting increase in reliability is well justified.

There are four main types of repairs:

- Routine wear
- Accidental breakage or damage
- Repairs due to operator neglect
- Routine overhauls

It pays to get machines ready ahead of time. Make all necessary repairs and have machines adjusted and ready to go before they are needed.

MM16633,000260A -19-11NOV10-2/2

Deciding When to Trade

In the early 1990s, cost of farm machinery steadily increased without a corresponding increase in prices for farm products. Longer machinery ownership, with increased emphasis on maintenance and repairs, helps to compensate for the problem of the cost-price squeeze.

Fortunately, farm machinery is built to last. Records show combines lasting well beyond 3,000 hours and tractors as much as 12,000 hours. Except for extremely high annual use, most machines do not reach the point of minimum cost before the 15th year. However, if machinery is kept too long, repair costs increase and reliability is lost.

Establishing Trading Guidelines

Good guidelines are important for making management decisions on when to trade machinery. Five important reasons for trading a machine are:

- The accumulated average annual cost per unit of use has reached its lowest point and is increasing
- The machine is obsolete in comparison to new models
- The machine has lost its reliability, meaning it is no longer dependable
- The machine is worn out
- An increase in the farm size makes a machine too small for timely operations

In all cases, investment credit or other tax credits need to be included when considering the purchase of new machinery. It is impossible to address these issues in this book due to the constant changes in tax laws. Instead it is suggested that you first learn how to make an accurate cost analysis, and then determine how it is affected by tax laws. When you do this, you will have a strong foundation on which you can make decisions.

For most machines, the earliest feasible trade time occurs at approximately three-fourths of the least-cost life. Certain machines may have an earlier feasible trade time, depending on the rate of usage. When buying a machine

Carefully compute best trade-in time to minimize expenses and maximize profit

that is a new concept, use a shorter obsolescence life of about 6 years.

MM16633,000260B -19-09DEC10-1/1

Considering Future Capacity Needs

An important aspect of buying a new machine or tractor is making the decision based on the machine size needed. Buying a tractor or machine with adequate power or capacity is a good idea, even if financing the purchase is necessary.

If decisions are based on available cash, a tractor purchased could end up being used only 100 hours per year, or an undersized tractor may require too much time to complete a job.

Another potential basis for machine selection is to purchase equipment large enough to allow one person to do all of the work. This might be the best choice for smaller, less diversified operations.

Chances are, however, that managing larger, more diversified farming operations requires more than one person to perform several ongoing jobs. The involvement of hired labor on expensive machinery usually means some type of supervision and the need for the manager to be available for on-the-spot decisions.

The balance between labor and machine size is an important one. Certain types of agricultural enterprises

Buy the size that will get important jobs done on time

may require a large amount of hired labor, while labor requirements on other types may be reduced by larger machines. The important thing is that labor-saving dollars might be eliminated or surpassed by high fixed costs of large machines. Consider these factors when selecting machinery or tractor size.

An integral concept is tractor or machine size selection regarding future increases in the size of an agricultural operation.

MM16633,000260C -19-15OCT10-1/1

Calculating Custom Work Costs

Hiring custom operators is one important alternative to owning machinery. In some cases, using custom operators completes the work faster, provides the least-cost method, and does not require the capital needed for owning a machine. In other cases, doing additional custom work can help a farm operator justify ownership costs.

When considering hiring a custom operator, do not forget timeliness considerations. Waiting for a custom operator to arrive is expensive in terms of timeliness if it means not getting crops planted or harvested at the optimum time. Always consider timeliness in making a decision to use either a custom operator or an alternative method.

For some operations, it may be better to use a custom operator

Continued on next page   MM16633,000260D -19-15OCT10-1/2

Determining and Comparing Costs

Determining when to use a custom operator is one of the most important decisions made in machinery management. There is a rather simple formula for calculating the break-even point for owning machinery versus the cost of custom work.

The formula for calculating the break-even point is:

$$\text{Break-Even Point} = \frac{\text{Average Annual Fixed Cost}}{(\text{Custom Rate/Unit Area} - \text{Operating Costs/Unit Area})}$$

Fixed costs are incurred by owning a machine, regardless of annual use. In the case of the custom operator, these fixed costs must be recovered with the custom charge. Operating costs, which include costs for fuel, lubricants, labor, and repairs, must also be included in the custom charge.

A custom operator receives additional income from each acre (or hectare) to pay off the annual fixed costs.

For small annual use, a custom operator may provide the least-cost method

A— **Average Cost per Acre**
B— **Cost to Own**
C— **Custom Rate**
D— **Acres (Hectares) per Year**

MM16633,000260D -19-15OCT10-2/2

Decision Time — Selecting the Best Alternative

Good machinery management is a constant process of evaluating alternatives. Limiting the options for consideration is a common mistake. Personal prejudice, pride, and sometimes available capital can narrow alternatives and preclude what may well be the best answer. The overriding goal is to keep costs at a minimum while achieving profit goals.

What are the alternatives?

If your tractor is 12 years old, has 6,600 hours, and needs extensive repairs, it's decision time. The alternatives include:

- Rent
- Lease
- Repair and keep on owning
- Buy a new tractor of the same size
- Sell it and make due with the other tractors

Continually evaluate alternatives

- Buy a new tractor of a different size
- Buy a used tractor

Cost of owning and operating the equipment is the final criterion for decision-making, providing it meets the operational needs.

MM16633,000260E -19-15OCT10-1/1

Case Study

Financial Analysis

A financial analysis is a process used to determine the health of a farm or ranch business. The success of the farm manager often depends more on the use of a financial analysis than any other management tool. For this reason, financial statements must be considered as an integral part of farm and ranch management. A financial statement will do the following:

- Help obtain credit
- Determine net worth of the business and which way it is going
- Determine liquidity — the ability to convert assets to cash
- Determine solvency — measures the amount left after converting assets to cash and paying all debts
- Determine equity, particularly the ratio of debt to equity, another measure of solvency
- Determine borrowing power at any time

A financial analysis shows where you have been and where you are headed, financially. In other words, it can measure the success of the business.

Peer Pressure

Rick Young is 27 years old and has been lucky to get an early start in farming on his own. He took over from his father when he was 19 years old. He is an only child, has been married for 3 years to Jane, and loves to farm.

At first, he made the usual mistakes of a young, eager farmer. He worked hard, learned from his experiences, and soon gained a reputation for being a good farmer. He has a good mix of new and used equipment with plenty of capacity to get everything done on time. Within a few years he expanded his operation to 1,500 acres, mainly as a tenant. It seems that everyone wants him for a tenant. Besides 400 acres of his parents' land, he bought 90 acres with a farmstead, including an old house.

By using soil tests, quality seed, and proper chemicals, Rick's yields slowly increased. His family has always been conservative, and Rick continues that tradition. Instead of building a new house, he and Jane remodeled the old farmhouse on the land that they purchased. With the same attitude, Rick is always on the lookout for bargains in used equipment. He tries to keep his equipment inventory low in value but adequate to provide plenty of capacity in the field.

Rick does not belong to an outside record-keeping service, but he has a filing system where he can retrieve any information necessary for tax purposes or for making decisions. Increasing yields coupled with shrewd marketing and low machinery costs mean that he is in good shape financially. Other than loans for a storage bin and a combine, his only major debt is a mortgage on the 90 acres where the farmstead is located.

In the past, Rick has made most of his own decisions after consulting with family members. Today he is being influenced more and more by his neighbor Steve Sample. Steve is about his age and started farming about the same time. Steve is heavily in debt. He recently bought a new four-wheel drive tractor and a new house. Steve convinces Rick that he also should get a four-wheel drive tractor. He convinces Rick that it would save him time because of the added capacity. Rick likes the idea of more leisure time for him and Jane.

Rick first talks to his local dealer and finds that he can trade for the new tractor for $80,000 and his two older tractors. Rick next visits his lending institution. His lender, George Scott, is 60 years old. He has lived through the economic decline of the 1980s. George says, "Rick, I have known your family for a long time. I also know that you would qualify for the loan, but let's start with a financial statement just to see how you are doing." Rick works out the financial statement. He sees that his net worth has been steadily increasing. His cash flow is better than the average. Projections for the next 2 years show that he will be able to meet all operating costs and existing loan payments if prices remain at the level of the past 2 years.

The new loan of $80,000 for 6 years would mean an annual payment of $18,000. George says, "While the $80,000 loan would still keep your equity-to-debt ratio within allowable limits, you would be left without much borrowing power for emergencies. If you were 30 years older and your business stabilized, then I would recommend that you ask for the loan. It would make life easier, but it would do very little, if anything, to increase your net income. You are still in a growth process and need to use your capital for productive assets. You also need a reserve for emergencies. If prices decline still further, you could be in serious trouble within 2 years."

The more George talks, the more Rick realizes what George is trying to tell him. He really doesn't need the big tractor. It would be used only 250 hours a year and the cost per acre would be higher than necessary. As a result, Rick trades the oldest tractor, which is 12 years old and needs extensive transmission work, for a new 150-horsepower tractor. This is about half the cost of the four-wheel drive tractor that he first thought he needed. Table 3 shows why the decision to buy the smaller tractor kept Rick and Jane in a much better financial position.

Available Borrowing Power	
Current	$90,000
After Trading for Four-Wheel Drive Tractor	$10,000
After Trading for Two-Wheel Drive Tractor	$50,000
Annual Loan Payments	
Four-Wheel Drive, 6 Year Loan	$18,000
Two-Wheel Drive, 6 Year Loan	$9,000
Table 3 — Case Study	

Table 3 — Case Study

Continued on next page

MM16633,000260F -19-11NOV10-1/2

Analysis: Rick's situation is typical. He starts conservatively and stays out of heavy debt. Then as time goes on and he gains confidence and experience, he feels that he must keep up with his neighbors. George's advice is sound and reaffirms the need for him to deal with his own situation and not be influenced by his neighbors.

By trading for the smaller tractor and continuing to work longer hours, he saves $9,000 a year. He also maintains a better reserve in borrowing power for emergencies. The bottom line is the fact that the four-wheel drive tractor would not increase income and may place Rick in a precarious financial position.

MM16633,000260F -19-11NOV10-2/2

Glossary of Terms

Definitions of Terms (A—S)

A

ABRASIVE — Material that cuts or grinds.

ABSORBED — To be taken into or bound tightly to another compound or element.

ABSORPTION — Process of being "sucked up" or taken into; specifically, entry of a chemical into an organism such as a plant or insect.

ACARICIDE — Pesticide used to control mites and ticks.

ACID-FORMING — Materials which leave an acid residue in the soil.

ACID SOIL — Soil with pH less than 7.0.

ACRE — An area of land equal to 43,560 sq. ft.

ACTIVE INGREDIENT — That part of a pesticide formulation directly responsible for the pesticidal effect.

ACTIVE SENSING SYSTEMS — Sensing systems that generate a signal, bounce it off an object, and measure the reflected signal.

ACTUATOR — A device used in variable-rate application that responds to controller signals to regulate the amount of material applied to a field.

AEROSOL — Very fine particles of a pesticide suspended in air.

AEROSOL CAN — Container used to apply pesticide that is stored under pressure and driven through a fine nozzle.

AGGREGATE — A unit of soil structure consisting of primary soil particles (sand, silt, clay) bonded together.

AGGREGATED STRUCTURE — A soil that consists primarily of soil particles which are bonded together. The particles are bonded together by natural forces and substances derived from organic matter and microbial activity.

ALGORITHM — An ordered set of rules or instructions written as a computer program designed to assist in finding a solution to a problem. For example, an algorithm can be created to permit a microprocessor to relate sensor input to actuator output onboard a crop chemical applicator.

AMMONIA — A gas made up of 82.25% nitrogen, 17.75% hydrogen, and carries a positive charge. It is not stable at normal pressures and temperatures and thus is lost to the atmosphere through volatilization if not incorporated into the soil. The chemical symbol is NH_3.

AMMONIUM — A compound containing a combination of nitrogen and hydrogen that carries a positive charge. The chemical symbol is NH_4.

AMMONIUM NITRATE — A nitrogen fertilizer material that contains ammonium and nitrate. The chemical formula is NH_4NO_3.

AMMONIUM SULFATE — A fertilizer compound that contains ammonium and sulfate ions. The chemical formula is $(NH_4)_2SO_4$.

ANALYSIS — The percentage composition of various nutrients in fertilizer.

ANHYDROUS AMMONIA — A nitrogen fertilizer material containing nitrogen and hydrogen. It is a gaseous compound at normal pressure and temperature. Therefore, it is kept in a pressurized vessel for storage and application. The terms ammonia and anhydrous ammonia are used interchangeably.

APPLICATION — Putting chemicals on or in soil or plants.

APPLICATION RATE — Amount of chemical applied per acre.

APPLICATOR — Device for applying chemicals, usually fertilizers or granular pesticides. A person who applies chemicals.

AQUATIC WEED — Weeds that grow in or near water.

AS-APPLIED MAP — A document or digital representation of position- and time-referenced crop production input applications within a field.

ATTRIBUTE — A numeric and/or text description of a spatial entity.

AUGER — Mechanism using spiral flights around a central cylinder to move harvested forage to a bale chamber or forage harvester cutterhead.

B

BAIT — Food or other material used to attract a pest to a trap or pesticide.

BALE — Forage compressed into a rectangular or cylindrical package to increase density for easier handling, transport or feeding.

BALER — Machine used to compress hay into rectangular or cylindrical package. See Rectangular Balers and Round Balers.

BALLAST — Weight added to tractor or implement to improve tractor traction and stability and/or improve penetration of soil working tool.

BAND APPLICATION — Application made in a narrow band, usually over or alongside a row.

BARN CURING — Artificial drying of hay with natural or heated air, usually to reduce exposure time in the field and possible weather damage.

BATCH-TYPE YIELD MONITOR — A yield monitor that weighs the amount of harvested grain as it sits in the combine grain tank or as it is being unloaded. Yield must be calculated using an estimate of the area harvested.

Continued on next page

"

BIENNIAL — Plant that matures and produces seed during the second year of its life and dies after not more than two years of life.

BIOLOGICAL CONTROL — Use of biological agents, such as parasites and predators, to control pests.

BOOM — Pipe or tubing with several nozzles to apply chemicals over a wide area at one time.

BREAKDOWN — An unexpected change in duty status from operational to non-operational, due to mechanical failure.

BROADCAST — To spread over the entire surface of the soil. A method of sowing seeds by randomly scattering them on the ground.

BROADCAST APPLICATION — Application made nonselectively over a broad area.

C

CALIBRATE — To determine the application rate of a sprayer or other application equipment.

CARBAMATE — A family of pesticides having similar chemical structures.

CARBON DIOXIDE — A noncombustible gas formed by combining carbon and oxygen.

CAUTION — Signal word used to alert users to slightly toxic pesticides.

CHANNEL — The bed of a stream, river, or other watery course. A natural stream that conveys water. A ditch excavated for the flow of water.

CHISEL PLOW — A primary tillage implement that breaks or shatters the soil, leaving it rough and with residue on or near the surface. Operating depth generally ranges from 6 to 12 inches (15 to 30 cm). The basic frame has two to four cross-members to which staggered curved shanks are attached. The effective shank spacing is usually 12 to 15 inches (30 to 38 cm). Interchangeable soil engaging components including various types of sweeps, spikes and shovels are attached to the shanks. Spikes and sweeps do less soil mixing and cover less residue than twisted shovels. In heavy or wet residue, the chisel plow may clog, thus it is common practice to shred the residue or disk before chiseling. Some chisel plows are equipped with a gang of coulters or disk blades mounted in front to cut the residue. This gang reduces the need for an operation before chiseling and when used as a single operation leaves more residue on the surface than disking prior to chiseling. A gang of coulters in front of the chisel plow leaves more residue on the surface than a gang of disks.

CHLOROSIS — Loss of green color (yellowing or whitening) in foliage.

CLAY — Soil with particles smaller than 0.0001 inches (0.002 mm) in diameter; plastic when moist, hard when dry.

CLEAN TILLAGE — A sequence of tillage operations that produces a soil surface having essentially no plant residues or growing vegetation except the particular crop desired.

CLIMATE — The kind of weather a place has over a period of years. Climate includes conditions of heat and cold, moisture and dryness, clearness and cloudiness, wind and calm.

CLOD — A small clump of soil.

COLLECT-AND-WEIGH — A method for determining crop yield, typically on a whole-field basis. Each truck or wagon load of grain is weighed as it leaves the field and the moisture content is determined by sampling the load.

COMBINATION TILLAGE IMPLEMENT — Consists of a wide variety of components commonly found on other tillage tools. For example, a combination implement may have two single-acting disk gangs in front, followed by three or four rows of field cultivator shanks and shovels, followed by a multi-row spike-tooth harrow. Other components on combination implements are rotary cutter reels (rolling stalk choppers) and rolling baskets. Combination implements are often used for one-pass incorporation of chemicals. Incorporation has been reported to be uniform horizontally and to a shallower depth than provided by two passes of a tandem disk or field cultivator. Most combination implements will operate in heavy residue conditions without clogging.

COMPACTION — See SOIL COMPACTION.

CONCENTRATE — A pesticide, as sold, before being diluted for application.

CONCENTRATION — Amount of active ingredient in a formulation or mixture.

CONDITIONING — Crushing, cracking and/or bruising of plant stems soon after they are cut to hasten drying.

CONDITIONING ROLLS — Smooth (rubber or steel) or corrugated (rubber or molded urethane) rolls on mower-conditioners or windrowers to condition plants for faster drying.

CONSERVATION — Protection and preservation of natural resources.

CONSERVATION TILLAGE — A sequence of tillage operations which leave some or much residue on soil surface (usually at least 30% cover) when preparing soil for planting. Objectives are reduced soil erosion, moisture retention in soil, and saving of fuel, time, and labor.

CONTACT — To touch or be touched by.

CONTAMINATE — To accidentally get a pesticide into other pesticides, seed, fertilizer, food, and feed.

CONTAMINATION — The introduction into water of microorganisms, chemical, organic or inorganic wastes, or sewage, which renders the water unfit for its intended use.

CONTROL — To reduce the number or growth of pests in an area.

Continued on next page

MM16633,0002574 -19-09NOV10-2/11

CONTROL SEGMENT — The portion of GPS consisting of a network of monitoring stations used to update satellite navigation signals.

CONTROLLER — An electronic device used to change product application rates on-the-go.

CONVENTIONAL TILLAGE — Sequence of tillage operations traditionally or most commonly used in a specific geographic area. A tillage method which includes plowing, disking and harrowing prior to planting or seeding.

CONVEYOR CHAIN — A chain mechanism which conveys or carries plants to the opened furrow.

COPPER SULFATE — Most common source of copper in fertilizer.

CORM — A thick, rounded subsurface stem base with leaves and buds attached, which acts as a vegetative reproductive structure.

COULTERS — Sharp steel disks, usually flat, used to cut trash and define furrow slice ahead of moldboard plow bottoms, leaving clean furrow wall and reducing wear on share and shin; to provide lateral stability for rear-mounted row-crop cultivators and disk tillers; or assembled in rows and attached to the front of chisel plows (stubble mulch tillers) or field cultivators to cut through residue and reduce plugging. May have plain, notched, or ripple-edge blades, and some are concave rather than flat.

COVER CROP — A close growing crop grown primarily for the purpose of protecting and improving soil between periods of regular crop production. Cover crops are also used to protect new seedlings and grown between trees and vines in orchards or vineyards. A legume, such as alfalfa or clover, is sometimes used as a cover crop, as it will also provide some of the nitrogen needed for the following grain crop.

CROP RESIDUE — The portion of a plant, or crop, left in the field after harvest. Usually refers only to the portion of the plant material above the soil surface.

CROP RESIDUE MANAGEMENT — The system used in handling plant residue. It begins with the selection of crops grown and includes all aspects of how a crop is grown, and harvested and how the residues are handled after harvest. Including all field operations that affect the amount of residue on the soil surface.

CROP ROTATION — The growing of different crops in recurring succession on the same land.

CUSTOM RATE — The rate of charge for a custom operation.

CUTTERBAR — Cutting device with a reciprocating knife having triangular blades or sections or with rotary disks to cut standing crops.

D

DEFOLIANT — Chemical that removes leaves.

DENITRIFICATION — A microbial process that results in the conversion of nitrates to gaseous nitrogen compounds that will be lost when soils are saturated with water.

DEPARTMENT OF DEFENSE (DOD) — The organization responsible for the creation and operation of the Global Positioning System.

DEPOSIT — Pesticide left on leaves, stems, fruit, other plant parts, or skin after pesticide application.

DESICCANT — Chemical that draws moisture out of a plant, causing it to wither and die.

DETACHMENT — A taking apart; separation. Rain and wind can cause detachment of soil particles.

DETERIORATE — Break down or decay.

DICOTYLEDON — Plants including legumes, cotton, and tobacco having two cotyledons or seed leaves.

DIELECTRIC — A material that does not conduct electricity. Examples include plastic and dry grain.

DIFFERENTIAL CORRECTION — Correction of a GPS signal to improve its accuracy. The correction is performed using a second stationary GPS receiver positioned at a known location. The second receiver computes the error in the signal by comparing the true distance from the satellites to the GPS measured distance.

DIFFERENTIAL GLOBAL POSITIONING SYSTEM (DGPS) — A method of using GPS that improves the position accuracy through differential correction.

DIGITIZE — To digitally record the relative position of a point, line, or area located on a map.

DISEASE — Change in one or more physiological processes of a plant or animal that impairs its ability to use energy.

DISK BLADE — Concave, spherical, or conical ground-driven rotating blades. The blades are mounted on primary and secondary tillage implements, shallow row-crop cultivators, and other implements.

DISPOSAL — Discarding a pesticide or pesticide container.

DOLOMITIC LIMESTONE — A liming material that contains both calcium carbonate and magnesium carbonate.

DOMESTIC ANIMAL — Tame animal, such as a cow, sheep, or dog (livestock and pets).

DRAWBAR PULL — The force exerted by the power unit at the drawbar to pull an implement.

E

ECOLOGY — Study of the interrelationships between plants, animals, and their surroundings.

Continued on next page

MM16633,0002574 -19-09NOV10-3/11

EFFECTIVE FIELD CAPACITY — Actual work accomplished (acres or hectares per hour) by an implement despite loss of time from field-end turns, inadequate tractor capacity, deficient tractor or implement preparation, adverse soil conditions, irregular field contours, lack of operator skill, or other factors. See THEORETICAL FIELD CAPACITY, FIELD EFFICIENCY.

ELECTROMAGNETIC ENERGY — Energy that is reflected or emitted from objects in the form of electrical and magnetic waves that can travel through space.

ELECTROMAGNETIC SPECTRUM — All wavelengths of electromagnetic energy including X-rays, ultraviolet rays, visible light, infrared light, microwaves, and radio waves.

ELEVATOR — Machine with endless chain-and-slat or belt mechanism used to move bales, usually to a higher elevation such as a barn loft.

EMERGENCE — The point in plant growth when the plant actually breaks through the soil surface.

END OF PASS DELAY — A delay that allows any grain that passes by the flow sensor after the combine header has been raised to be included in yield calculations.

END-WHEEL DRILL — Drill is supported by wheels at each end that are used to drive the seeding mechanism.

ENVIRONMENT — All of the surrounding things, conditions, and influences affecting the growth or development of living things.

EPHEMERAL GULLY — A channel formed by flowing water. The size of the channel is larger than a channel formed by rill erosion but smaller than channels formed by gully erosion.

EROSION — The detachment and movement of soil or rock by running water, wind, ice, or other geological agents, including the force of gravity.

EXPOSE — To come into contact with a pesticide.

EXPOSURE — State of having been exposed.

F

FAILURE — The inability of a machine to perform its function under specified field and crop conditions.

FALLOW — Allowing cropland to lie idle either tilled or untilled during the whole or greater portion of the growing season. Tillage or herbicides are used to control weeds. The practice is used to increase the storage of moisture in the soil for the next crop.

FEATURE — A geographic component of the earth's surface that has both spatial and attribute data associated with it. Examples include a field, well, or waterway.

FERMENTATION — The chemical change occurring in silage and haylage during storage.

FERTILIZATION — The addition of essential nutrients as either chemical compounds or as organic materials.

FERTILIZER — Material that contains one or more plant-food elements to be incorporated into the soil to provide nutrients for the plants. Commercial fertilizers include the major nutrients: nitrogen, phosphorus, and potash.

FIELD — A specific location on a farm that is given a distinct designation such as "North Forty."

FIELD CAPACITY — The maximum amount of water a soil can hold after free drainage.

FIELD CULTIVATOR — Similar to a chisel plow, but is lighter in construction and designed for less severe conditions. Field cultivators generally have three or four ranks of equally spaced flexible shanks. Either C-shaped or S-shaped shanks are used. The shanks are typically spaced 24 to 40 inches (61 to 100 cm) apart on each rank of a 4-rank machine providing an effective spacing of 3 to 7 inches (7.6 to 17.8 cm). Soil-engaging components available for field cultivators include reversible points or shovels and sweeps. Points or shovels are 1-1/2 to 2 inches (3.8 to 5 cm) wide, and sweeps vary in width from 4-1/2 to 12 inches (11.4 to 30 cm). Field cultivators operate at depths of 3 to 5 inches (7.6 to 12.7 cm). Field cultivators are used for primary tillage where surface residue is light and soil conditions permit. Usually, however, field cultivators are used for secondary tillage of previously tilled soil, especially for incorporating herbicides.

FIELD CURING — Permitting hay or forage to dry naturally in a swath or windrow prior to chopping or packaging.

FIELD EFFICIENCY — Ratio of effective field capacity to theoretical field capacity. Ratio of the area an implement can theoretically cover in one hour and the area actually covered. Ratio is expressed as a percentage of theoretical field capacity.

FIELD TIME — The time a machine spends in the field measured from the start of functional activity to the time the functional activity in the field is completed.

FIXATION — Process of rendering available plant nutrients unavailable or fixed in the soil.

FIXED CHAMBER ROUND BALER — Round baler design that produces only one size of bale.

FIXED COSTS — Costs that do not depend on the amount of machine use, such as depreciation, interest on investment, taxes, insurance, and storage.

FLAIL — Cutting device with swinging knives or blades on a rotating horizontal shaft. Material is usually cut into several pieces as it is struck by succeeding flails and carried from standing position to the discharge point. Flails are used in different machine configurations to mow plants, chop forage, or as pickup units for forage harvesters and stack wagons.

FLOTATION — Ability of tractor or implement tires to stay on top of soil surface; usually related to soil condition, tractor or implement weight, wheel slippage, and contact area between tires and soil surfaces.

Continued on next page

MM16633,0002574 -19-09NOV10-4/11

FLOW SENSOR — A sensor that measures the amount of material flowing through a conduit per unit of time.

FOGGER — Aerosol generator.

FOLIAGE — Leaves, stems, and other above-ground plant parts.

FORAGE — Hay, grass, grain crop, or other food for animals. Usually considered to be a solid seeded crop (drilled) as compared to a row crop.

FORAGE BLOWER — A fan-type conveyor used for placing chopped forage into storage structures.

FORAGE HARVESTER — Machine that chops hay or forage into short lengths for easy storage or handling. With different header attachments it can cut standing crops, pick up windrows, snap ear corn, or gather stover.

FORMULA — Shorthand method of expressing the constituents of a compound with symbols and abbreviations.

FORMULATION — Mixture of one or more pesticides and other materials such as diluents and carriers as sold — does not include adjuvants or tank mixes added at the time of application.

FUMIGANT — Volatile chemical that kills pests with a gas or vapor.

FUNGI — Small plant organisms that lack chlorophyll and cause rots, mildews, and other diseases (singular: fungus).

FUNGICIDE — Pesticide used to kill or control fungi that cause plant diseases.

FURROW OPENER — Device that opens the soil and shapes the furrow into which the seed is dropped.

G

GAUGE WHEEL — Wheel used to adjust height of furrow opener to control the depth of the furrow.

GEOGRAPHIC DATA — Data that contains not only the attribute being monitored but also the spatial location of the attribute. Also known as spatial data.

GEOGRAPHIC INFORMATION SYSTEM (GIS) — A system, usually computer-based, for the input, storage, retrieval, analysis, and display of geographic data. A GIS database is usually composed of map-like spatial representations called layers or coverages. These layers (coverages) may contain information on a number of attributes including land elevation, land use, land ownership, crop yield, and soil nutrient levels.

GEOREFERENCED DATA — Spatial data that pertains to a location on the earth's surface.

GEOREFERENCING — The process of associating non-spatial data such as crop yield values with geographic coordinate data to produce spatial data.

GERMINATION — The point at which a seed having adequate moisture, temperature, oxygen, and light begins to grow.

GLOBAL POSITIONING SYSTEM (GPS) — A network of satellites controlled by the U.S. Department of Defense designed to help determine a radio receiver's position in latitude, longitude, and altitude. GPS is not a synonym for precision farming; GPS is only one technology that is used in precision farming.

GRASS — A member of the botanical family Gramineae, characterized by blade-like leaves arranged on the culm or stem in two ranks.

GRID CENTER METHOD — Soil sampling method in which samples are taken from the center of a grid cell. Also known as grid point sampling or point sampling.

GRID SAMPLING — Soil sampling method in which a field is divided into square sections (grids) of several acres or less. Samples are then taken from each section and analyzed.

GROUND COVER — Any vegetation producing a protecting mat on or just above the soil surface. In forestry, low growing shrubs and herbaceous plants under trees.

GROUNDWATER — Water found underground in porous soil and rock strata.

GROWTH REGULATORS — Chemicals that alter the normal growth or reproduction of a plant.

GULLY EROSION — The process whereby flowing water accumulates in narrow channels and removes soil from the bottom and sides of the channels. The depth of the channels can range from 1.5 feet (0.5 m) to as much as 100 feet (30 m).

GYPSUM — A material that contains calcium and sulfate. The chemical symbol is $CaSO_4$.

H

HARROW — An implement used to level the soil surface, redistribute surface residue to enhance moisture retention, pulverize clods, and disturb germination of weeds. Common soil-engaging components on harrows are spring-steel shanks with a variety of points, rigid square, diamond, or round teeth, round coil-tine teeth, and chain drags. The harrow frame may be rigid or flexible. Harrows are often attached to the rear of disk harrows, field cultivators, or drills to smooth and firm the soil surface and redistribute residue.

HARVEST AID — A chemical (usually a desiccant or defoliant) that prepares a crop for harvest.

HAYLAGE — Low-moisture silage with usually about 40% to 50% moisture content.

HAZARDS — Risk of danger to applicator, observers, livestock, and desirable plants.

Continued on next page   MM16633,0002574 -19-09NOV10-5/11

HERBICIDE — Pesticide used to control weeds and unwanted plants. Chemical substance, either granular or liquid, applied to the soil to kill the roots of broadleaf and grassy weeds. See PESTICIDE.

HOPPER — A container from which the contents can be emptied slowly and evenly.

I

ILLEGAL RESIDUE — Quantity of pesticide on the harvested crop which exceeds the legal limit.

IMMOBILE NUTRIENTS — Nutrients that do not move down through the soil with the movement of water.

IMPACT PLATE — A plate placed in the path of grain flow. The force with which the grain strikes the plate is measured and used to estimate grain flow rate.

IMPERVIOUS SOIL — A soil that is resistant to penetration by water and usually by air and roots.

IMPLEMENT — A piece of equipment used for farming. Chisel plows, planters, and combines are implements.

INFESTATION — Damaging pest population.

INFILTRATION — The flow of water into a soil through pores or other openings.

INGEST — Eat or swallow.

INJECT — Force pesticide or fertilizer into a plant or the soil.

INSECTICIDE — Pesticide used to control or prevent damage by insects. A substance that destroys insects by chemical action.

INSOLUBLE — Not soluble; will not dissolve. As applied to phosphorus in fertilizer it means that portion of the total phosphorus which is neither soluble in water nor neutral ammonium citrate. As applied to potash and nitrogen it means not soluble in water.

INSTANTANEOUS YIELD MONITOR — A yield monitor that continuously measures and records crop yields on-the-go.

INTEGRATED PEST MANAGEMENT (IPM) — A controlled approach that blends cultural and chemical practices to keep pests from reaching economic injury levels.

INVERSION PRIMARY TILLAGE — Primary tillage that partially or completely inverts the soil to bury residue from the previous crop.

IRRIGATION — The application of water to soil to assist in the production of crops.

K

KILOGRAM (Kg) — A Metric unit of weight approximately equal to 2.2 pounds.

KNIVES — Shallow-working tools for killing weeds close to such crops as beets, beans, and vegetables with minimum surface disturbance.

L

LABEL — Printed information attached to a pesticide container which contains important directions and limitations regarding applications, pests controlled, and crops for which its use has been approved.

LAND — The total natural and cultural environment within which plant growth takes place; a broader term than soil. In addition to soil, its attributes include other physical conditions, such as mineral deposits, climate, and water supply; location in relation to centers of commerce, populations, and other land, the size of the individual tracts or holdings; and existing plant cover, works of improvement, and the like.

LAND TENURE — The length of time that an individual will control a piece of land.

LARVA — Immature or wormlike (grub) stage of an insect that does not resemble the adult form (plural: larvae).

LEACHING — Removing nutrients from soil by passage of water through the soil profile. The process of removing soluble salts from soil by passage of water through the soil. This is a primary step in the improvement of saline soils.

LEASE — An contract that provides a method of obtaining the services of a piece of equipment in return for periodic payment. The period of a lease is normally considered as longer than one year.

LETHAL — Toxic or deadly.

LIME — Lime from the strictly chemical standpoint refers to only one compound, namely calcium oxide (CaO). However, the term lime is commonly used in agriculture to include a great variety of materials which are usually composed of the oxide, hydroxide, or carbonate of calcium or of calcium and magnesium. The most commonly used forms of agricultural lime are ground limestone (carbonates), hydrated lime (hydroxides), burnt lime (oxides), marl and oyster shells. A material whose calcium and/or magnesium content is used to neutralize soil acidity.

LINE-TRANSECT — A method of measuring the percentage of the soil surface covered with plant residues.

LIQUID FERTILIZER — Any fertilizer formulated as a liquid for handling and application.

LITER (L) — Metric unit of volume equal to approximately 1.06 quarts.

M

MAINTENANCE — Cleaning, oiling, greasing, adjusting, etc., to keep a machine in operative condition and to help maintain its efficiency.

Continued on next page

MANAGEMENT UNIT — An area or subunit of a farm field that has a functionally homogeneous combination of yield-limiting factors for which a single rate of a specific crop production input is appropriate. There can be a different set of management units for each type of input or treatment that a field receives.

MANURE — The waste of animals. It may contain straw or other materials used as an absorbent, or it may be straight feces and urine. Animal waste used as fertilizer.

MATERIAL — A substance; a word often used here to refer to pesticides.

MECHANICAL AGITATOR — Device that mixes spray materials in a tank by mechanical means.

MICRONUTRIENT — Trace elements or minor nutrients; materials needed by plants in very small quantities. One of the seven essential elements found in low concentrations in plants. The elements are boron, chlorine, copper, iron, manganese, molybdenum, and zinc.

MICROORGANISM — An animal or vegetable organism too small to be seen except with a microscope. Bacteria are microorganisms.

MINIMUM TILLAGE — Tillage where one or more tillage processes are left out. A system of conservation and mulch tillage using the minimum number of operations necessary to produce an acceptable yield. May involve combined or reduced operations. Also called optimum, reduced, and economy tillage.

MITE — Tiny animal with eight legs, often mistaken for or referred to as an insect.

MOISTURE CONTENT — Percent of total product weight which is water. Moisture content equals 100 minus percent dry matter content of the product.

MOLD — Fungus with conspicuous, profuse, or woolly growth.

MOLDBOARD PLOW — A primary tillage implement which cuts, partially or completely inverts a layer of soil, and thereby covers, buries, and mixes surface materials, and pulverizes the soil. The bottom or base and one side of the plow cuts the soil. The moldboard is the curved plate above the bottom, which receives the slice of soil and inverts it. Moldboard plows are equipped with one or more bottoms of various cutting widths. Two-way moldboard plows are equipped with right-hand and left-hand moldboards that are alternately used to turn all slices in the same direction as the plow is operated back and forth across the field.

MONOCOTYLEDON — Plants including all grasses having one cotyledon or seed leaf.

MOWER—Machine used to cut standing vegetation.

MOWER-CONDITIONER — Combination mower and conditioner that has conditioning rolls approximately the same length as the cutterbar.

MULCH—A natural or artificial layer of plant residue or other materials on a soil surface. A mulch conserves moisture, holds soil in place, may aid in establishing plant cover, and minimizes temperature fluctuations.

MULCH TILLAGE — Tillage or preparation of soil in such a way that plant residues are left on the surface.

MULCH-TILL — A conservation tillage system which includes the use of implements that till the entire soil surface. At least 30% of the soil surface must be covered with plant residues after planting.

N

N — Symbol for nitrogen.

NATURAL ENEMIES — Predators and parasites (insects or diseases) that attack pests.

NAVSTAR — NAVigation by Satellite Timing And Ranging.

NECROTIC — A browning or dying of plant tissue.

NEMATICIDE — Pesticide used to control nematodes.

NEMATODE — Tiny, tubular, unsegmented, eel-like worms that feed on plant roots.

NERVOUS SYSTEM — Brain, spinal cord, and nerves of humans and animals.

NITRATE — A compound containing a combination of nitrogen and oxygen that carries a negative charge. The chemical symbol is NO_3.

NITRIFICATION — Bacterial conversion of ammonic nitrogen to nitrate nitrogen. The microbial conversion of ammonium to a nitrite and then to a nitrate.

NITRIFICATION INHIBITORS — A chemical compound that inhibits the organisms that convert ammonium to nitrite.

NITROGEN (N) — An inert gas that makes up about four-fifths of the air. Nitrogen for commercial purposes can be "fixed" synthetically from the atmosphere by several processes. A nutrient critical to plant growth.

NOXIOUS WEED — A plant defined by law as particularly undesirable.

O

OPERATING COSTS — Costs that depend directly on the amount of machine use.

OPERATING SPEED — Steady ground speed of a machine (miles or kilometers per hour).

OPERATION DELAY — The time required for grain to move from the combine header to the grain flow sensor.

ORAL — Through the mouth.

ORGANIC — Material largely composed of carbon.

Continued on next page

MM16633,0002574 -19-09NOV10-7/11

ORGANIC MATTER — Decomposed, partially decomposed, or un-decomposed plant or animal material. Material composed largely of carbon, including carbon of decayed plant material or of artificial or synthetic origin. Decomposed plant residue component of the soil mass. Important in aggregation of soil particles (soil structure), moisture absorption and fertility of soil.

ORGANIC NITROGEN — Nitrogen compounds that contain carbon. Examples of materials that contain organic nitrogen include organic matter, crop residues, animal manures, etc.

ORGANIC SOIL — Soil high in organic matter. A soil that contains a high percentage (greater than 15%) of organic carbon throughout the upper and most weathered part of the soil profile.

ORGANISM — Any living thing: plants, animals, bacteria, insects, and fungi.

ORGANOPHOSPHATE — A group of similar insecticides, including malathion, parathion, and diazinon.

ORNAMENTAL — Plant grown for beauty, accent, color, and screening.

P

P — Symbol for phosphorus.

PARASITE — A plant, insect, or animal that lives and feeds on or in a living host plant, insect, or animal.

PASSIVE SENSING SYSTEMS — Sensing systems that measure naturally-emitted and reflected signals.

PASTURE — A field or hillside on which cattle, sheep, or horses, or other grazing animals can feed.

PATHOGEN — An organism or agent capable of causing disease.

PEAT — Soil material consisting primarily of raw undecayed or slightly decomposed organic matter.

PERENNIAL — A plant that normally lives more than two years.

PEST — An unwanted organism, such as a weed, insect, or virus.

PESTICIDE — A chemical or physical agent used to kill or control pests. Any one of various substances used to control harmful insects (insecticide), fungi (fungicide), weeds (herbicide), or other living organisms that destroy or inhibit plant growth or are otherwise harmful.

pH — A term used to indicate the degree of acidity or alkalinity. A material that has a pH of 7.0 is neutral. Values above 7.0 denote alkalinity and below 7.0 denote acidity. Chemically, pH is the negative logarithm of the hydrogen ion concentration.

PHEROMONE — Chemicals that insects produce in order to communicate with each other.

PHOSPHORUS (P) — One of the 16 chemical elements that is required by plants for normal growth and development. A highly-reactive element that combines readily with other elements and is one of the three primary plant foods.

PHOTOSENSOR — A sensor that is used to detect light.

PHOTOSYNTHESIS — The basic process of plant life, by which carbon dioxide and water, in the presence of sunlight and nutrients, are converted by chlorophyll to organic substances, with liberation of oxygen.

PHYTOTOXIC — Injurious to plants.

PLANT DISEASE — An abnormal plant condition caused by a pathogen, virus, or improper environmental condition.

PLANT NUTRIENTS — Elements required by plants for adequate growth and development.

PLANTING DEPTH — The depth of the furrow in the soil made by the openers to accept the seed when it is dropped. The depth is usually controlled by adjusting the gauge wheels. Adjusting the press wheel or the pressure on the openers are other methods that have been used to adjust the planting depth.

PLOW PAN — See PLOW SOLE.

PLOW SOLE — Compacted layer, restricting root and water movement, which may form in some soils just below the tilled area after several years of primary tillage to the same depth.

POLLUTION — The presence in a body of water (or soil or air) of substances of such character and in such quantities that the natural quality of the body of water (or soil or air) is degraded so as to impair its usefulness or render it offensive to the senses.

PORE SPACE — The small openings, holes, or spaces between soil particles that are filled with gases, air, or liquids (water).

POSITIONING SYSTEM — A general system for identifying and recording, often electronically, the location of an object or person.

POTASH (POTASSIUM OXIDE) (K_2O) — The potassium content of fertilizers is expressed as potash.

POTASSIUM (K) — A highly-reactive element that combines readily with oxygen and many anions.

POTENTIOMETER — A device that produces a changing electrical resistance as the relative positions of its components are changed.

POWER HOP — Simultaneous loss of tractor traction and a bouncing, pitching ride. Occurs most frequently under high drawbar loads in certain soil conditions when using mechanical front wheel or four-wheel drive tractors.

PRECAUTIONS — Safety measures taken in advance.

Continued on next page

MM16633,0002574 -19-09NOV10-8/11

PRECISION FARMING — Managing each crop production input – fertilizer, limestone, herbicide, insecticide, seed, etc. – on a site-specific basis to reduce waste, increase profits, and maintain the quality of the environment.

PREDATOR — An animal or insect that feeds on and destroys other animals or insects.

PRESS WHEEL — Wheel used to firm soil around the seed to make good seed-to-soil contact. Has also been used on some planters to gauge planting depth and to drive the seeding mechanism.

PRESS-WHEEL DRILL — Drill has press-wheel gangs to firm the soil over the seed. The press-wheel gangs drive the seeding mechanism.

PRESSURE — Force on an area, usually expressed as pounds per square inch; pressure causes a liquid to flow.

PRESSURE SENSOR — A sensor that produces an electrical signal proportional to a fluid pressure.

PRIMARY NUTRIENTS — The three essential elements of nitrogen, phosphorus, and potassium are classified as primary nutrients as they are required in relatively large amounts and are the ones that most often limit crop production.

PRIMARY TILLAGE — The deepest tillage operation used in a tillage operation. Primary tillage implements include moldboard, chisel, and disk plows; heavy tandem, offset and one-way disks; and subsoilers.

PROPERTIES — Characteristics or traits used to describe something, such as a chemical.

R

RADAR (RADIO DETECTION AND RANGING) — A method of determining the position or velocity of an object by bouncing high frequency signals off the object and measuring the reflected signal.

RADIOMETRIC SYSTEM — A yield monitoring system that determines mass flow rate by measuring the reduction in the intensity of a radioactive stream of particles as the grain obstructs the flow of the radioactive particles.

RATE — Amount of material applied to a plant or an area of land.

RECTANGULAR BALER — Machine used to compress hay into rectangular packages. Conventional balers use a reciprocating plunger to form bales which may be tied with two or three wires or twines. Large rectangular balers form bales weighing up to 2,000 pounds (908 kg) and wrapped with four to six plastic or twine ties.

REDUCED TILLAGE — Refers to any system that is less intensive and less aggressive than conventional tillage. Compared to conventional tillage, either the number of operations is decreased or a tillage implement that requires less energy per unit area replaces an implement typically used in the conventional system. The term reduced tillage is not very meaningful because of geographical differences.

REGISTRATION — Acknowledgement of approval of a pesticide, by EPA, for the uses stated on the label.

RELIABILITY — The ability of a machine to perform at its intended quality and capacity level during the time period that it is scheduled to operate.

REMOTE SENSING — The act of detection and/or identification of an object, series of objects, or landscape without having the sensor in direct contact with the object.

RENT — Similar to a lease, except the term of the contract is usually a year or less, for example, a day, hour, week, or month.

REPAIR — Restoring a machine to operative condition after breakdown, wear, accident, etc. Repairs do not add to the value or prolong the life of a machine. Major overhauls are not repairs.

REPELLENT OR REPELLANT — Chemical which discourages insects or other pests from entering a treated area.

RESIDUE — Pesticide remaining on or in a plant or other treated area following a time lapse after the application.

RESIDUE MANAGEMENT — See CROP RESIDUE MANAGEMENT.

RESTRICTED-USE PESTICIDES — Pesticides which may be used only by licensed or certified applicators.

RILL EROSION — A small intermittent watercourse with steep sides, usually only a few inches (mm) deep and, hence, no obstacle to tillage operations.

RODENTICIDE — Pesticide used to control rats, mice, and other rodents.

ROOT ZONE — The top portion of a soil profile in which plant roots grow.

ROTARY HOE — Consists of one or two staggered gangs of spider-like wheels spaced about 3-1/2 to 4 inches (8.9 to 10 cm) apart. A rotary hoe is a fast, economical way to control small weeds and break surface crust to improve crop emergence. Operating speed is usually 6 to 10 mph (9.6 to 16.1 km/hr); the draft requirement is low. Rotary hoes, especially those with self-cleaning provisions, can be used in most conservative tillage systems.

ROTARY MOWER — Machine with one or more rotating blades, commonly used to cut weeds, stalks, grass, and brush.

Continued on next page

MM16633,0002574 -19-09NOV10-9/11

ROW-CROP CULTIVATOR — An implement used after a row crop has emerged to till the soil for increased aeration and water infiltration and to kill weeds. Several types of row cultivators are available including: 1. S-tine or Danish-tine designed to operate at shallow depths and high speed in tilled soil without much residue. 2. C-shank (multiple shank) equipped with three to five shanks per row. Various sweep shapes and sizes are available and can also be equipped with weeding disks. The C-shank cultivator has good soil penetration, is usually operated at a slow speed (2-1/2 to 4 mph (4 to 10 km/hr)) but often clogs when used in heavy residue. 3. C-shank (single shank) handles surface residue found in conservation tillage situations. A coulter to cut the residue is usually mounted in front of each shank. Each row assembly consists of a shank, with a 16 to 24 inch (41 to 61 cm) wide sweep and two weeding disk blades (or disk hillers).

ROUND BALER — Machine that rolls hay or other forage in a cylindrical bale chamber to form bales weighing 1,000 to 2,000 pounds (454 to 908 kg). The bales may be tied with twine or surface wrapped or left untied.

RUNOFF — Flow of surface liquid from land, plant, or animal surfaces.

S

SECONDARY NUTRIENTS — The secondary plant foods include calcium, magnesium and sulfur. Less-critical elements required in smaller amounts for plant growth than nitrogen, potassium and phosphorous.

SECONDARY TILLAGE — The tillage operations used to kill weeds, cut and cover crop residues, incorporate herbicides and prepare a seedbed. Some secondary tillage tools are light- and medium-weight disk harrows, field cultivators, row cultivators, rotary hoes, and drag harrows.

SEED SPACING — The distance between seeds in the row. Adjusted by changing the combination of drive and driven sprockets to change the ratio between metering mechanism and forward travel speed.

SENSOR-BASED VARIABLE RATE APPLICATION SYSTEM — A system that adjusts product application rate on-the-go based on information received from real-time sensors.

SERIAL PORT — A connector on a computer which can be used to communicate to other serial devices such as a modem. Serial refers to the protocol used for the communications. The most common serial protocol is RS-232.

SHANKS — Standards extending down from implement frames to which soil-working points, shovels, and sweeps of row-crop cultivators and other tillage implements are attached; may be rigid, flexible, or spring-cushioned.

SHOVELS — Deep-working (compared to sweeps and knives) soil-working tools for row-crop and field cultivators, chisel plows, and other implements.

SHEET EROSION — The movement of soil by a uniform sheet of flowing water.

SILAGE — Green forage converted to animal feed through fermentation. Normal moisture content is 65% to 70%.

SILO — Structure for storing silage or haylage-may be a vertical cylinder or a horizontal trench or bunker.

SILT—A soil consisting of particles between 0.002 and 0.00008 in. (0.05 and 0.002 mm) in equivalent diameter. A soil textural class.

SITE-SPECIFIC CROP MANAGEMENT (SSCM) — The use of variability of soil and crop parameters to make decisions on the application of production inputs.

SITE-SPECIFIC YIELD MAP — A representation of field crop yields collected on-the-go by a harvester equipped with an instantaneous yield monitor. Each location/site in a field is assigned a specific crop yield value.

SOIL — The mineral and organic material developed from weathering mineral and decaying organic matter. It covers most of the immediate surface of the earth and serves as a natural medium for the growth of plants.

SOIL COMPACTION — Process of moving and rearranging soil particles to decrease pore space and increase bulk density. Usually detrimental to crops, but may be beneficial in promoting seedling emergence.

SOIL CONSERVATION SERVICE (SCS) — A service organization provided by the United States Department of Agriculture (USDA). The primary goal of the service is soil conservation.

SOIL FUMIGANT — A pesticide applied as a vapor or gas to the soil subsurface.

SOIL pH — A numerical measure of the acidity or hydrogen ion activity of soil. It is used to determine lime requirement. A pH of 7.0 is consider to be neutral. Values less than 7.0 are consider to be acid and values higher than 7.0 are consider alkaline. See pH.

SOIL PROFILE — A vertical section of the soil from the surface through all horizons.

SOIL RESISTANCE — The resistance of an implement as it is moved over or through the soil to accomplish the desired results.

SOIL SAMPLE — A small amount of soil collected for nutrient analysis.

SOIL STRUCTURE — The combination or arrangement of primary soil particles into secondary particles or units.

SOIL TESTING — The analysis of a soil sample to determine the need for supplemental nutrient application.

SOIL TEXTURE — The physical structure or character of the soil determined by the relative proportions of the soil components (sand, silt, and clay) of which it is composed.

Continued on next page   MM16633,0002574 -19-09NOV10-10/11

SOIL TYPE — A term used to refer to the combination of primary physical constituents of a soil. For example, silty clay loam, fine sandy loam, or heavy clay.

SOLUTION — Mixture of solid, liquid, or gas dissolved in a liquid.

SPACE SEGMENT — The portion of GPS consisting of NAVSTAR satellites orbiting the earth at 20,200 km.

SPATIAL DATA — See GEOGRAPHIC DATA.

SPATIAL VARIABILITY — Differences in field conditions, such as plant, soil, or environmental characteristics from one location in a field to another.

SPECIES — A group of living organisms with similar characteristics.

SPEED SENSORS — Sensors that measure the rotational speed of a shaft or the reflection of radio or sound waves off the ground to determine machine speed.

SPLASH EROSION — The detachment and movement in air of soil particles caused by the impact of raindrops on a soil surface.

SPOT TREATMENT — Application to a small or limited area.

SPRAY — Application of a pesticide using water or another liquid as a carrier and applied in tiny droplets.

START OF PASS DELAY — A delay that allows the initial flow of grain before full flow is achieved to be ignored in yield calculations when starting a pass.

STARTER FERTILIZER — The placement of fertilizer near the seed at planting time, so named because it tends to give an early season growth advantage.

STATIC WEIGHT SPLIT — Relative weight distribution on front and rear axles of a tractor.

STRAIN GAGE — A device that has a changing electrical resistance as it is deformed. Used in load cells to convert force to electrical signals.

SUBSOIL — The soil below the tilled soil (or its equivalent of surface soil), in which roots normally grow. Although a common term, it cannot be defined accurately. It has been carried over from early days when "topsoil" was conceived as the plowed soil and that under it as the "subsoil."

SUBSOILER — A primary tillage implement, somewhat similar to a chisel plow. It is typically designed to operate 12 to 22 inches (30 to 56 cm) deep. Subsoilers are used primarily to alleviate soil compaction, therefore, used when the soil is dry for maximum effectiveness.

SURFACE WATER — Water located aboveground.

SURFACE WRAP — Wrapping round bales with mesh or plastic sheet rather than twine.

SUSCEPTIBLE — Organism that can be killed or injured by a pesticide.

SUSTAINABLE AGRICULTURE — Crop production system that emphasizes cultural practices and minimizes pesticides. Also called low-input sustainable agriculture and regenerative agriculture.

SWATH — Width of the area treated in one pass by a sprayer or other applicator.

SWEEPS — Soil-working tools with wide cutting edges for cultivators, chisel plows, stubble-mulch plows and other tillage implements.

SYMBIOTIC — The living together of two organisms in a mutually beneficial relationship. Example; rhizobium bacteria that provide nitrogen to legumes and in turn gain their nutrition from the legume plant.

SYNERGISM — An increase in toxicity of chemicals mixed together, beyond their total potency as separate agents.

MM16633,0002574 -19-09NOV10-11/11

Definitions of Terms (T—Z)

T

TANKAGE — Dried animal residues, usually from fat or gelatin.

TARGET — Area or plant intended to be treated.

TERRACE — An embankment or combination of an embankment and channel across a slope to control erosion by diverting or storing surface runoff instead of permitting it to flow uninterrupted down the slope.

THEORETICAL FIELD CAPACITY — Rate of performance obtained if a machine performs its function 100% of the time at a given operating speed using 100% of its theoretical width.

TILLAGE — Mechanical soil-stirring actions for nuturing crops by providing suitable soil environment for seed germination, root growth, and weed and moisture control.

TILLAGE OPERATION — A single pass over the soil with equipment designed to till the soil.

TILLAGE SYSTEM — The sequence of tillage operations performed in producing a crop.

TILLING SEEDER — Tillage implement that falls between disk and moldboard plow in function. Consists of spherical blades mounted on a common axial shaft that prepare a seedbed for each seed drop on the drill. Normally throws soil to the right. Considered mostly as a dryland implement.

TILTH — Physical condition of the soil and its potential for cultivation of growing plants.

TIMELINESS — Ability to perform an activity at such a time that quality and quantity of product are maximized.

TOOLBARS AND TOOL CARRIERS — Basic frames for assembling do-it-yourself tillage and planting implements from scores of component parts, available from implement dealers, including gauge and transport wheels, standards and shanks, soil-engaging tools, hitches, braces, clamps, markers, hydraulic cylinders and hoses, unit planters, and others.

TOPOGRAPHY — The surface features of a field or region. The topography includes hills, valleys, streams, lakes, bridges, tunnels, and roads.

TOPSOIL — The uppermost part of soil, ordinarily disturbed by tillage, or its equivalent in uncultivated soil. The depth often ranges from 6 to 8 inches (15 to 20 mm) and is usually higher in organic matter than the material below.

TOTAL COSTS — The sum of fixed and operating costs.

TOXIC — Poisonous.

TOXICANT — A poison.

TOXICITY — Degree to which a substance is poisonous.

TOXIN — "Poison" produced by a plant or animal.

TRACTION — Effective force resulting from thrust of tractor tires against soil or other surface; depends on such factors as nature of surface, contact area between tires and surface, and tractor power and weight.

TRANSPORTATION — The action of carrying from one place to another. Flowing water or wind can transport soil particles long distances.

U

UREA — A synthetic organic-nitrogenous compound often used as a fertilizer. A fertilizer that contains carbon, nitrogen, hydrogen, and oxygen. The chemical formula is $CO(NH_2)_2$.

USDA — The United States Department of Agriculture.

USEFUL LIFE — The service life of a machine before it becomes unprofitable for its original purpose due to obsolescence or wear.

USER SEGMENT — The portion of GPS consisting of receivers used by civilians and the military for determining the position of a person or object.

V

VARIABLE-RATE APPLICATION (VRA) — Adjustment of the amount of cropping inputs such as seed, fertilizer, and pesticides to match conditions in a field.

VARIABLE-RATE TECHNOLOGY (VRT) — The equipment used to perform variable-rate application.

W

WHEEL RAKE — Rake that forms windrows using revolving wheels with teeth moving through the swath.

WHEEL SLIP — Reduced forward speed of tractor drive wheels resulting from soil conditions that limit total traction. Excessive slippage is detrimental, but some slippage is desirable to cushion tractor engine and drive train from sudden overload.

WIDE RANGE — Capability of a pesticide to control a variety of pests.

WIND EROSION — The detachment, transportation and deposition of soil by the action of wind.

WINDROW — Row of hay or forage formed by a rake, mower-conditioner or windrower to be picked up later by a harvesting machine.

WINDROWER — Machine for cutting forage or grain in which material is carried to the center of the table or platform by an auger or conveyor belts and fed into a narrow conditioner or dropped onto the ground.

Y

Continued on next page

MM16633,000258E -19-04OCT10-1/2

YIELD LIMITING FACTOR — The plant, soil, or
environmental characteristic or condition that keeps a crop
from reaching its full yield potential within any specific
area in a farm field.

YIELD MAP — A representation of crop yields collected
on-the-go by a harvester equipped with an instantaneous
yield monitor. Each location/site (pixel) in a field is
assigned a specific crop yield value.

YIELD POTENTIAL — The yield expected for the
particular soil type under a given level of management.

Z

ZERO-TILL PLANTING — Planting in soil undisturbed
from previous crop except for tillage provided by planter
or drill.

MM16633,000258E -19-04OCT10-2/2

Index

Continued on next page